T0079808

INFERENCES WITH IGNORANCE: LOGICS OF QUESTIONS

INFERENTIAL EROTETIC LOGIC & EROTETIC EPISTEMIC LOGIC

MICHAL PELIŠ

CHARLES UNIVERSITY IN PRAGUE
KAROLINUM PRESS 2016

KAROLINUM PRESS
Karolinum Press is the publishing department of Charles University in Prague
www.karolinum.cz

Designed by Jan Šerých
Typeset by studio Lacerta (www.sazba.cz)
Printed in the Czech Republic by Karolinum Press, Prague
First edition

Cataloging-in-Publication Data is available from the National Library of the Czech Republic

ISBN 978-80-246-3181-3
ISBN 978-80-246-3193-6 (pdf)

The manuscript was reviewed by Mariusz Urbański (Adam Mickiewicz University in Poznań)
and Igor Sedlár (Comenius University in Bratislava)

CONTENTS

PREFACE

As the title indicates, the book is about logic of questions (erotetic logic) which is a branch of non-classical logic. The text goes in for two formal (logical) systems of questions with the important role of erotetic inferential structures. The first one, *inferential erotetic logic*, had been originally developed in Poland in 1990s. I used this system in a slightly rearranged version that serves as the main inspiration for erotetic logic studied in the epistemic framework. The second system, *erotetic epistemic logic*, is just such a combination of epistemic logic with questions that is open for application in public announcement logic.

The text is based on my doctoral dissertation [36] which was finished during the year 2010. I decided to change some parts with respect to reviews and discussions with my colleagues and students. Nonetheless, the structure and main results are almost the same. I tried to incorporate all objections as well as the advice of my referees or, at least, I comment on them in the last chapter. The aim of this book is rather different from the thesis, therefore I checked allof the proofs and explanations and amended them. Although I believe that the text is more comprehensive now, some background knowledge of the reader is expected. A knowledge of elementary notions from formal (mathematical) logic and a basic knowledge of modal logic is assumed, and would be very helpful in reading the part starting at Chapter 3. Nice introductions to (modal) epistemic logic are introductory chapters in [48] and [11].

STRUCTURE OF THE BOOK

The book includes two main parts that can be read independently. The first one is Chapter 2 and the second one consists of Chapters 3 and 4. Chapter 1 serves as an introduction and provides common a methodology for both parts. The last Chapter 5 contains some final remarks on the methodology used, summarizes the main results, problems, related approaches, and also further directions of the branch.

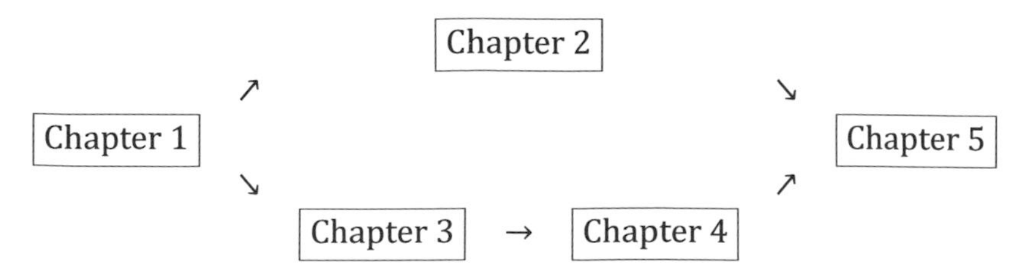

If we compare the contents of the chapters, we may find another division of the book. Both Chapter 2 and Chapter 3 can be understood as a study of the 'logic of questions'. There we are interested in basic erotetic structures, i.e., inferences with questions, relationships of questions and declaratives, and answerhood conditions. However, Chapter 4 introduces questions as a part of communication; a dynamic approach is applied in the erotetic epistemic framework introduced.[1]

CHAPTER 1: LOGIC AND QUESTIONS

The chapter introduces a multi-paradigmatic situation in the methodology of erotetic logic and contains a short historical overview of this branch with a special emphasis on recent developments.[2] We briefly introduce *inferential erotetic logic*, Groenendijk-Stokhof's intensional approach, and some developments of these theories. However, the core of the chapter is devoted to a formalization of questions based on sets of answers. We justify the usefulness of the set-of-answers methodology in the study of erotetic consequence relations as well as in an epistemic interpretation of questions.

This chapter is based on the paper [37].

CHAPTER 2: CONSEQUENCE RELATIONS IN INFERENTIAL EROTETIC LOGIC

This part is aimed at the studying of relationships among consequence relations that are introduced in *inferential erotetic logic* (IEL). We keep the framework of IEL, but the question representation uses the methodology from Chapter 1. IEL requires that declarative and interrogative formulas are not mixed on the object-language level; answers are strictly declarative sentences. The defined consequence relations with questions are naturally based on the multiple-conclusion entailment among sets of declarative formulas. We add the term *semantic range* of a question to the terminology of IEL and work with sets of declaratives as associated with classes of models. This 'model-based approach' makes proofs and properties very transparent. The chapter is a technical overview of some IEL concepts and their properties. We understand the IEL presented as providing a general framework and inspiration for the work with inferences among questions and declaratives.

Chapter 2 can be read as a full introduction to the topic; no reading of another text is required. The chapter was originally published in [35].

---

1 Dynamic-like approaches bear the name 'logic of inquiry' in literature.

2 Approximately till the year 2010. Comments on the newest literature are in Chapter 5.

CHAPTER 3: EPISTEMIC LOGIC WITH QUESTIONS

The main goal here is to incorporate questions in a general epistemic framework. Questions are represented by finite sets of direct answers and their 'satisfiability' in a state of an epistemic model is based on three conditions that express ignorance and the presuppositions of a questioner.[3] In the framework of normal modal logic, a question becomes a complex modal formula. Inspired by inferential structures in IEL, we show that there are 'philosophically' similar structures based on a classical implication. The rest of the chapter is devoted to answerhood conditions and the role of the implication with respect to the epistemic context and conjunctions of yes-no questions.

This chapter cannot be read without a basic knowledge of modal (epistemic) logic.

CHAPTER 4: A STEP TOWARDS THE DYNAMIZATION OF EROTETIC LOGIC

This chapter takes full advantage of the multi-agent extension of the setting from Chapter 3 and can be considered an application of the introduced erotetic-epistemic approach in a dynamic framework. We define here *public announcement logic* based on the S5 modal system extended by group modalities. The askablity of questions as well as answerhood conditions are studied from the viewpoint of groups of agents. As an application, we show the role of questions and group modalities in a (communication-like) answer 'mining' in a group of agents.

Some results of the chapter were published in [38, 39].

CHAPTER 5: CONCLUSION

The last chapter includes a brief overview of related works, main results, possibilities of future directions, and comments on the problematic parts of our approach. This chapter is the result of many discussions. The important part of the subsection devoted to weak points and problems is based on the reviews to the original doctoral thesis. A list of up-to-date publications cannot be complete. However, we tried to add important papers closely related to our approach.

---

3 We will use the term *askability* of a question (for an agent in a state).

ACKNOWLEDGEMENTS

It would be a long list of names to express my thanks to everybody who assisted me in the work on erotetic logic. And it was many names and grants I listed in the preface to my doctoral dissertation [36]. Therefore, now, I will mention especially people, they helped me in the preparation of the book.

The great help, with an improvement of the text, came from my thesis referees Andrzej Wiśniewski and Igor Sedlár. They carefully reviewed my dissertation and gave me a lot of helpful comments and recommendations. Many of them were worked in. Andrzej as well as Igor supported me with their papers and advice. Igor together with Mariusz Urbański took on board the review of this book and helped me to prepare the final version. Thank you.

My special thanks go to Ondrej Majer who has helped me with the work in dynamization of erotetic epistemic logic. We wrote some papers from this field and, simultaneously, we have been working in epistemic relevant logic. We have been organizing seminars on dynamic logics for some years. The audience of the seminar was the best test of my ideas; let me mention particularly Marta Bílková, Michal Dančák, Adam Přenosil, Vít Punčochář, and Petr Švarný.

Last but not least, my thanks go to my colleagues and friends at the Faculty of Arts and at the Institute of Philosophy. Let me mention, especially, Lenka Jankovská, Paweł Łupkowski, Jaroslav Peregrin, Vladimír Svoboda, and Vítězslav Švejdar.

The work on this book was supported by the Institute of Philosophy of the Czech Academy of Sciences and by the Program for the Development of Fields of Study at Charles University, No. P13 Rationality in Human Sciences, sub-program 'Modern logic, their methods and applications'.

1. LOGIC AND QUESTIONS

1.1 QUESTIONS, ANSWERS, AND INFERENCES

In this chapter we wish to show that it is reasonable to consider questions as a part of logical study. In logic, declarative sentences usually have their formal (logical) counterparts and play an important role in argumentation. We often see logic to be primarily a study of inferences. Inferential structures are studied in formal systems, which can differ in the formalization of declaratives as well as in admitting or rejecting some principles.

We believe that dealing with questions in the logical framework will be justified if we show that questions can play an autonomous and important role in inferences. Perhaps this point may be considered as the most important to justify *logic of questions.*[1]

This introductory chapter provides a brief overview of the history as well as methodology in recent approaches to erotetic logic. However, the main aim is to concentrate on the methodology used in the rest of the book—we introduce and discuss a variant of the methodology based on sets of answers.

1 In this paper we use the term *logic of questions* in the same meaning as *erotetic logic*, discussions on both terms can be found in [19].

1.1.1 QUESTIONS AND ANSWERS

Let us imagine a group of three friends: Anne, Bill, and Catherine. Each of them has one card and nobody can see the cards of the others. One of the cards is the Joker and everybody knows this fact.[2] Then

> Who has the Joker?

is a reasonable sentence in this situation. We recognize it as an *interrogative sentence* because of its word order and the question mark. Moreover, interrogatives are often connected with intonation and interrogative pronunciation.

An interrogative sentence includes more—a pragmatic aspect. A question is a "request to an addressee to provide the speaker with certain information", this is an *interrogative speech act* [17, p. 1057]. Pragmatically oriented approaches emphasize the roles of a speaker and an addressee. The roles seem, on the one hand, to be outside of the proper meaning of interrogatives. On the other hand, they are often understood to be a crucial aspect in analyses of questions; this can be the reason why some logicians argue against some variants of the logic of questions.

No matter what our starting position is if we want to work with interrogatives in the framework of a formal system, we are obliged to decide the following two problems at least:

1. How to formalize questions?
2. What is the (formal) semantics of questions?

Reviewing the history of erotetic logic, there is no unique solution. There are many approaches to the formalization of questions and every approach varies according to what is considered important. Logic of questions is multiparadigmatic. David Harrah [18, pp. 25–26] illustrates it nicely with examples of so called 'meta-axioms'. He groups them into three sets according to the degree of acceptance by erotetic logicians:

1. The first group includes meta-axioms accepted in almost all systems. Harrah calls them *absolute axioms*. For example:
    (a) Every question has at least one partial answer.
    (b) (In systems with negation) For every statement P, there exists a question Q whose direct answers include P and the negation of P.
    (c) Every question Q has a presupposition P such that: P is a statement, and if Q has any true direct answer, then P is true.

---

2 In the epistemic setting we will expect more: the rules of a 'game' are *commonly known*, cf. Chapter 4.

2. The second group, *standard axioms*, is often accepted, but not in all systems:
    (a) Every question has at least one direct answer.
    (b) Every direct answer is a statement.
    (c) Every partial answer is implied by some direct answer.
    (d) Every question is expressed by at least one interrogative.
    (e) Each interrogative expresses exactly one question.
    (f) Given an interrogative I there is an effective method for determining the direct answers to the question expressed by I.
3. The last group is called *eccentric axioms*. The following examples of such axioms are accepted only in some interrogative systems:
    (a) If two questions have the same direct answers, then the two questions are identical.
    (b) Every question Q has a presupposition that is true just in case some direct answer to Q is true.

Let us notice the terminology, the difference between *interrogative (sentence)* and *question* has been just introduced by standard axioms. The first term mostly refers to the type of a sentence and the second one is a bit more complex. A question is expressed by an interrogative (sentence) and can be 'posed', 'asked', etc., cf. [19]. Although we use *interrogative* and *question* in the same meaning here, the term *interrogative sentence* is reserved for a natural-language sentence, if necessary.

What seems to be common to all approaches is that questions are something structured and closely connected with their answers. We hardly find a propositional theory where questions are unstructured as atomic propositions are. The meaning of a question is always closely connected to its *answerhood conditions*.

Since an answer to a question is often represented by a declarative, the natural starting point of many erotetic theories is a standard formal system for declaratives.

> "Any first-order language can be supplemented with a question-and-answer system" [51, p. 37].

This broadly accepted statement invites us to solve the problem with the formalization of questions together with their meaning. Questions' autonomy depends on the chosen solution. Andrzej Wiśniewski distinguishes two basic groups of erotetic theories: *reductionist* and *non-reductionist*. Roughly speaking, non-reductionism is characterized by the claim that questions "are not reducible to expressions of other syntactic categories" [51, p. 40], see Section 1.2.1 in this chapter, too. The boundary between both groups is fuzzy. Perhaps only pure pragmatically oriented approaches belong to the radical reductionism with a complete rejection of questions as a specific entity in formal logic.[3]

1.1.2 INFERENCES WITH QUESTIONS

Although there are discussions on whether it is necessary to work with questions as a new specific entity in a formal system, almost all theorists agree that questions play a specific role in inferences. Let us come back to our group of friends. The situation, where

> *either Anne has the Joker or Bill has the Joker or Catherine has the Joker,*

can raise the question

> Q: *Who has the Joker?*

A question is raised (inferred) from a declarative or from a set of declaratives. What would make this raising reasonable? 'Answerhood conditions' is the answer. Any 'reasonable' answer to the question Q is connected to the declarative context.

Another kind of inferential structure could be based on declaratives as well as questions among premises. For example, from

> Q: *Who has the Joker?*

and

> Γ: *The only person from London has the Joker.*

can be inferred the question

> Q_1: *Who is from London?*

The relationship of the inferred question Q_1 and the initial question Q is based on their answerhood conditions. An answer to Q can provide an answer to Q_1 with respect to the context Γ. Moreover, in this example, Q can be inferred from Q_1 with Γ as well. This shows that the relationship is dependent on various kinds of answerhood conditions and contexts.

Let us have Q remain the same, but the context is

> *A person from London has the Joker.*

---

3 An example of one of the radical approaches [43] is commented in [34].

If two persons are from London and we gain their names in an answer to Q_1, then we receive only a *partial answer* to Q.[4] If each of the friends (or nobody) is from London, an answer to Q_1 does not provide any help for the answering of Q. Then we can discuss the utility of an inferential relation between Q and Q_1 with respect to this context.

The role of answerhood conditions in inferences among questions is obvious in the following example: From any (complete) answer to

Who has the Joker?

we obtain a (complete) answer to the question

Does Anne have the Joker?

as well as to the questions

Does Bill have the Joker?

and

Does Catherine have the Joker?

Answerhood conditions of the previous three questions are *entailed* in the answerhood conditions of the question Q. They can be inferred from the answerhood conditions of Q. The question *'Does Anne have the Joker?'* is *entailed* by *'Who has the Joker?'*.

We have just presented inference-like structures with questions as dependent on answerhood conditions. Now, still faced with the problem at how to formalize the relationship of questions and answers, we will introduce a convenient solution based on a liberal set-of-answers methodology.

1.2 SET-OF-ANSWERS METHODOLOGY

We are going to solve the problem of the formal shape of questions simultaneously with the problem of the questions' semantics. The formalization of a question will be based on a (possibly infinite) set of 'specific' answers. Moreover, we attempt to show that such an approach can also satisfy some semantic and pragmatic requirements.

4 Informally, a *partial answer* does not completely answer a question, but it eliminates some of the possible (and complete) answers. The term *possible and complete answer* expresses that the answer 'completely answers' a question. Later on we will introduce formal definitions.

1.2.1 SEMANTICS OF QUESTIONS

Some theories do not admit that questions could have an independent meaning in logic. Questions are paraphrased by declarative sentences; in particular, the question *'Who has the Joker?'* may be then paraphrased by

I ask you who has the Joker.

Another way is the paraphrasing by epistemic-imperative sentences:

Bring it about that I know who has the Joker!

The adequacy of both paraphrases in capturing the complete meaning of a question is rather problematic.[5] One of the problems is that these approaches are forced to work with a questioner and an addressee already on the basic level of the questions' meaning. Of course, we expect to utilize the importance of a questioner and an addressee, but it should be a task for a chosen background system not for the general semantics of questions.[6] We understand pragmatic aspects as a higher level analysis.

Nuel Belnap formulated three methodological constraints on the meaning of questions which he used for classification and evaluation of erotetic theories:[7]

1. **Independence** Interrogatives are entitled to a meaning of their own.
2. **Equivalence** Interrogatives and their embedded forms are to be treated on a par.
3. **Answerhood** The meaning of an interrogative resides in its answerhood conditions.

The most important is the first requirement, which is the main sign of non-reductionist theories. To accept the *independence* requirement means that we are obliged to look for the specific semantics of questions. The *equivalence* requirement is closely related to a semantic entailment and is dependent on the chosen semantics. *Answerhood* requires that the meaning of questions is related to the meaning of answers. In addition, we can work with the idea that the semantics of answers forms a good background for the study of the meaning of interrogatives.

Approaches, which accepts that answers are crucial for the meaning of questions, are in compliance with the first postulate from the following list suggested by Charles Hamblin:[8]

---

5 See also [19] for other examples and references.

6 In chapters 3 and 4 we will follow this idea and the background system will be dynamic epistemic logic.

7 Belnap, N.D., 'Approaches to the semantics of questions in natural language. Part I', Pittsburgh, 1981. Cited from [16, p. 3–4].

8 Hamblin, C.L., 'Questions'. Australasian Journal of Philosophy, 36(3): 159–168, 1958. Cited from [19].

1. Knowing what counts as an answer is equivalent to knowing the question.
2. An answer to a question is a statement.
3. The possible answers to a question are an exhaustive set of mutually exclusive possibilities.

Each postulate may be argued against and a detailed discussion is available in [17]. However, according to David Harrah, adopting the first one is "the giant step toward formalization often called *set-of-answers methodology*" [19, section 2]. Although there is not only one kind of set-of-answers methodology (SAM, for short) in the literature, we will not make any survey here. In the next subsection we introduce an easy idea of a question representation by a set of *direct answers*.

1.2.2 SETS OF ANSWERS

Generally, without any context, the question *Who has the Joker?* can be answered by expressions of the following form:

Anne.
Anne has it.
Anne has the Joker.
Anne and Bill.
⋮
Batman has the Joker.
⋮
Your friends.
People at this table.
⋮
Nobody.
⋮
etc.

The question seems to be answered if a (complete) list of Joker owners is given. We can assume that answers are propositions; thus, the first three items in the list have the same meaning in the answering of the question.

From the viewpoint of propositional logic and in accordance with the first two of Hamblin's postulates, we can understand every question as closely connected with a set of (propositional) formulas—formalized answers.

Furthermore, we can receive some of the following responses to the same question:

Anne doesn't have the Joker.

or

I don't know who has the Joker.

The first one can be considered to be a *partial answer*; it removes some answers as false, in particular, all answers with Anne having the Joker.[9]

The second one appears to bear another kind of information; an addressee says to a questioner that she has the same problem and would ask the same question. (We will return to this topic in the last paragraph of Section 1.4.)

If we had decided to represent every question by a complete set of its answers, we would not always have a clear and useful formalization of questions. Let us return to our example of three friends with cards. Considering the context and the question *Who has the Joker?*, a questioner expects one of the following responses:

α: *Anne has the Joker.*
β: *Bill has the Joker.*
γ: *Catherine has the Joker.*

or a response that leads to one of the just mentioned. In fact, the question

Who has the Joker?

with respect to the context

Either Anne has the Joker or Bill has the Joker or Catherine has the Joker.

might be reformulated to

Q': *Who has the Joker: Anne, Bill, or Catherine?*

The answers α, β, and γ are understood as 'core' answers that form the meaning of the question Q'. We use the term *direct answers* for them. The sentence

δ: *Neither Anne nor Bill have the Joker.*

is an answer, from which the answer γ can be inferred thanks to the context. We call δ *complete answer* to Q'. Complete answers are 'solutions' of a question and the set of direct answers is a subset of the set of complete ones.

Our SAM is inspired by the syntactic representation of questions in *inferential erotetic logic* founded by Andrzej Wiśniewski.[10] We want to be very liberal and, thus, to represent questions as sets of formulas, which play the role

9 We can imagine a context when the answer is complete—only two players.

10 The best overview of questions' formalization in inferential erotetic logic is in the book [51, chapter 3]. See also the article [53] and the new book [59]. We will return to this system in Chapter 2.

of direct answers. A (general) declarative language $\mathcal{L}$ is extended only by curly brackets ($\{, \}$) and a question mark (?). A question can be represented by the following structure

$$?\{\alpha_1, \alpha_2, ...\}$$

where $\alpha_1, \alpha_2, ...$ are formulas of the extended language.

No wonder that we have to impose some restrictions on direct answers to keep their exclusive position. Such restrictions are mostly a combination of syntactic and semantic requirements. From the syntactic viewpoint and being inspired by the previous examples, we require that (1) formulas $\alpha_1, \alpha_2, ...$ form a set (they are syntactically distinct) and (2) a set of direct answers has at least two elements. Both restrictions introduce questions as 'tasks' with at least two distinct 'solutions'. Syntactical distinctness is a first step to the idea that direct answers form the 'core' of a questions' meaning. In semantics, we will require non-equivalence above that.

The most typical questions with only two direct answers are *yes-no questions*. The question

Does Anne have the Joker?

has, in fact, the following two direct answers:

Yes. (*Anne has the Joker.*)
No. (*Anne does not have the Joker.*)

Such a question will be identified with the form $?\{\alpha, \neg\alpha\}$ and shortened as $?\alpha$. A yes-no question is a variant of a *whether question*, where an answer is a choice from two possibilities. The role of negation is considered to be very important in SAM. Negation is always related to a background system and receiving $\neg\alpha$ can mean something different from 'it is not the case of α'. For example, it can express a form of compatibility with the information that 'it is not α'.[11]

1.3 SAM IN THE RECENT HISTORY OF EROTETIC LOGIC

The logic of questions has, maybe surprisingly, a long history. F. Cohen and R. Carnap seem to be the first authors attempting to formalize questions in a logical framework—their attempts date back to the 1920s [19, p. 3]. The first 'boom' of logical approaches to questions took place in the 1950s (Hamblin, Prior, Stahl) and continued in the 1960s (Åqvist, Harrah, Kubiński). The first comprehensive monograph on questions [1] brought into life many important

11 Compare it with an interpretation of negation and compatibility relation in substructural logics, cf. [26] for an epistemic example with motivations.

terms used in erotetic logic so far. The late 1970s gave birth to influential reductionist theories: Hintikka's epistemic-imperative approach and Tichý's approach based on his *transparent intensional logic* [43].

We are not going to present a complete survey of erotetic theories. The reader can find a comprehensive overview of the history of erotetic logic in [19]. Erotetic theories with the main influence in this field of study are described in [51, chapter 2] as well. Moreover, both papers provide a good introduction to the terminology used in logic of questions and cover enough of the history of erotetic logic till the 1990s. The period from the 1950s till 1990s is described in [18]. Mainly linguistic viewpoint with the detailed discussion about the semantics of questions and pragmatic approaches can be found in [17].

Most influential modern logics of questions with an important role in erotetic inferences are the

- *inferential erotetic logic* (IEL) of Andrzej Wiśniewski and
- intensional approach of Jeroen Groenendijk and Martin Stokhof.

Both theories were appeared fully developed in the 1990s and we consider them as giving birth to several approaches, some of which are still influential.

1.3.1 INFERENTIAL EROTETIC LOGIC

Wiśniewski's IEL is a complex system dealing with various interrogative inferential structures. The influence of Belnap's and Kubiński's works is apparent. Primarily, it is based on classical logic and a formalization of questions, which is very similar to the SAM introduced. Consequence relations among questions and declaratives are defined on the metalanguage level where the role of multiple-conclusion entailment is important. The advantage of IEL is its possible generalization for non-classical logics. Chapter 2 is devoted to the slightly modified IEL. All important concepts concerning interrogative inferences are introduced and studied in their mutual relationships. In this chapter, the reader can find a list of relevant publications on this topic. The books [51] and [59], and the article [53] contain a nicely written presentation of Wiśniewski's approach.

The complex study of erotetic inferential structures in IEL predetermines an approach based on an old idea that a (principal) question can be answered by asking auxiliary questions. This kind of searching process can be seen as a tree with a principal question (and a context expressed by declaratives) in the root. Nodes bear auxiliary questions (with a context). The move from node to node is justified by IEL inferences in the direction of the leaves with answers.[12]

A similar idea is developed in Hintikka's *interrogative model of inquiry* reflecting the usefulness of questions in reasoning [21]. Hintikka's approach

12 See the paper [54] about *erotetic search scenarios.*

has an application in the game-theoretic framework for *belief revision theory*, cf. [12].[13]

The IEL methodology expanded to sequent style calculus, *socratic proofs*, and makes it possible to consider a derivation in terms of question-answer process; see [56] for classical propositional logic and [24] for some normal modal propositional logic.

1.3.2 INTENSIONAL EROTETIC LOGIC

Groenendijk's and Stokhof's approach might be called *intensional erotetic logic*. The meaning of a declarative sentence is given by truth conditions. For simplicity, let us imagine *logical space*, which is understood as a set of all 'possible' *states* (in the literature often called 'possible worlds', 'indexes' or 'situations'). Intension of a declarative is then a set of states where the declarative is true. Extensions are relative to states and the extension of a declarative in a given state is the truth-value of the declarative in the state.

Combining intensional semantics with a full acceptance of Hamblin's postulates we obtain the meaning of a question as a *partitioning* of logical space. In accordance with the third postulate, answers to a question form an exhaustive set of mutually exclusive propositions. Partitioning of logical space is the intension of a question. Extension of a question in a given state is the answer, which is true there. This is very similar to the SAM introduced in IEL. Both approaches accept the first two Hamblin's postulates, but the validity of the third one is accepted just in intensional erotetic logic.

Similarly to IEL, we can naturally define many important terms: partial answers, complete answers, informative value of answers, and entailment relation between questions. In the context of SAM it is interesting to mention the definition of entailment relation:

> A question Q *entails* a question Q_1 if and only if each answer to Q implies an answer to Q_1.

It means that the question Q provides a refinement of the partitioning given by the question Q_1. See [16] and [17].

This intensional approach has influenced many works in recent years. It is a good inspiration for epistemic representations of questions, see [6] and [33]. Very recently, the logic of questions has received more attention in connection with the dynamic aspects of epistemic logic and communication theory, cf. [46].

Groenendijk's and Stokhof's intensional interpretation inspired some extensional approaches. In paper [29], Gentzen-style calculus is presented and

13 IEL is presented as an alternative to Hintikka's approach (cf. [53, 54]).

paper [31] brings a many-valued interpretation of declaratives and interrogatives based on bilattices.[14] The work [40] gives a syntactic characterization of answerhood for the partition semantics of questions; on the top of that, the authors implement partition semantics in a question answering algorithm based on a tableaux theorem proving, cf. [41].[15]

The presented intensional approach deals with questions as having 'crisp' answers. However, we can imagine that there is a scale of answers, e.g., in case of yes-no questions, the scale

yes — rather yes — rather no — no

is usual in questionnaires. Such a kind of scale does not require to us introduce four new answers, but it corresponds to the comparative degrees of truth of the original yes-/no-answer. Truth-degrees are studied in multi-valued logics. The paper [4] presents propositional Groenendijk-Stokhof's erotetic logic with fuzzy intensional semantics based on fuzzy class theory. Although fuzzy logic seems to be suitable for the study of reasoning under vagueness, its combination with the logic of questions is still rather underdeveloped.

1.4 EPISTEMIC ASPECTS OF SAM

It is very natural to see a question as expressing the ignorance of a questioner delivered to an addressee. We mention it as a pragmatic aspect of questions. Looking for an epistemic counterpart of questions in the history of erotetic logic, the most known are *epistemic-imperative* approaches of Åqvist and Hintikka. These theories are reductionist ones, questions are translated into epistemic-imperative statements. Let us remember Hintikka's analysis [20] and the paraphrase *'Bring it about that I know who has the Joker!'* for the question *'Who has the Joker?'*. Both approaches are based on the idea of a questioner who does not know any answer to a question and who calls for a completion of knowledge.[16]

We have said that the pragmatic aspect of questions does not appear on the basic level of the questions' meaning analysis (Section 1.2.1). Now we are going to introduce an application of set-of-answers methodology in a general epistemic framework. It opens the door for the formalization of communication with questions inside dynamic epistemic logics. Epistemic analysis of questions has then two important parts. The first one (individual) works with the knowledge and ignorance of a questioner. The second one (group) considers an exchange of information in a group of agents.

---

14 An algebraic approach, where sets of answers form distributive lattices, is also studied in [22].
15 Nice comments are in [14].
16 More details are in [19] and [51, chapter 2].

Let us return to the example with the cards. If, in particular, Catherine wants to find out where the Joker is, then she can ask

Who has the Joker: Anne, or Bill?

From the set-of-answers viewpoint, the question has a two-element set of direct answers:

α: *Anne has the Joker.*
β: *Bill has the Joker.*

The question is 'reasonable' in this situation. By asking it, Catherine expresses that she

1. does not know the right answer to the question,
2. considers the answers α and β to be possible, and, moreover,
3. presupposes what is implicitly included in the set of direct answers $\{\alpha, \beta\}$, i.e., that either Anne has the Joker or Bill has it.

SAM makes it possible to specify expectations and presuppositions of a questioner. A questioner expresses not only the ignorance concerning the Joker's holder, but the presupposition that the holder must be either Anne or Bill. The set of direct answers informs what answers are considered as possible.

To employ the epistemic aspects of a question we have to indicate a questioner and an addressee. In accordance with the thesis that epistemic aspects are up to the background system, we enrich the language of any multi-agent propositional epistemic logic by curly brackets and indexed question mark ($?_x$), where x is the 'name' of an agent or of a group of agents—x indicates the questioner(s). In particular, Catherine's question can be formalized by

$$?_c\{\alpha, \beta\}$$

The role of an addressee is then a task for dynamic versions of epistemic logic. For example, in the framework of public announcement logic, which will be studied in Chapter 4, Catherine asks the question publicly among her friends and the addressee is the whole group of friends-agents {Anne, Bill, Catherine}.

Asking (publicly) the question $?_c\{\alpha, \beta\}$ the addressee can obtain the following information:

1. The agent c does not know whether α or β.
2. The agent c considers α and β as her epistemic possibilities.
3. The agent c expects a complete answer leading either to α or to β.

If c is not a liar or the question is not a rhetorical one, these items can be understood as delivering a good picture of c's epistemic state to an addressee. It is important for cooperative communication that an agent fully reveals her ignorance as well as expectations. A publicly asked question can then solve a problem (without being answered). Let us suppose that Anne has the Joker and Catherine has just asked the question $?_c\{\alpha, \beta\}$. Then Bill can infer:

> Catherine doesn't know who has got the Joker and she expects that either I have got it or Anne has got it. I haven't got the Joker. Therefore Anne must have it.

Catherine's publicly asked question was *informative* for Bill. He had the same problem *Who has the Joker?*, which is now solved. The formal definition of informative questions in the framework of public announcement logic will be studied in Chapter 4 (see Section 4.3.2).

1.4.1 *I DON'T KNOW* ANSWER

A question is often understood as a 'problem to solve'. Whenever we ask a question Q and someone answers

> *I don't know,*

we can interpret this answer as he or she has 'the same' problem, he or she would ask 'the same' question. In fact, we received an answer to a question

> *Would you ask the question Q?*

Liberal set-of-answers methodology makes it possible to deal with such questions; questions can be among direct answers as well.

In our example, whenever Bill wants to find out whether *Who has the Joker?* is a task for Catherine, he could ask

$$?_b\{?_c\{\alpha, \beta\}, \neg ?_c\{\alpha, \beta\}\}$$

It is a yes-no question where the first direct answer means that Catherine *does not know* who has the Joker or, rather, that she would ask the question formalized by $?_c\{\alpha, \beta\}$. The second one means that Catherine would not ask the question $?_c\{\alpha, \beta\}$. The reason can be that she knows a complete or a partial answer or she does not expect an answer just from the set $\{\alpha, \beta\}$.

As we have seen, Bill can use such 'questions about questions' to obtain information without revealing his own ignorance about the Joker-card holder. The context where each agent knows what card he or she holds, creates a situation where the interrogative sentence *Who has the Joker?* can have a specific form for each agent in SAM; in particular, $?_c\{\alpha, \beta\}$ for Catherine and $?_b\{\alpha, \gamma\}$ for Bill. These questions would not correspond to a situation where not one of them knows his or her own card. Then the question *'Who has the Joker?'*, in the SAM formalization $?_G\{\alpha, \beta, \gamma\}$, is a problem for each member of the group of agents

G = {Anne, Bill, Catherine}. The formalization of questions in SAM can help to reveal the ignorance as well as knowledge structure of an agent (group of agents) in a given context to addressees.

Most of the points mentioned in this section will be studied in detail in Chapters 3 and 4.

2. CONSEQUENCE RELATIONS IN INFERENTIAL EROTETIC LOGIC

2.1 INTRODUCTION

The inferential structures that will be introduced and studied in this chapter are based on the slightly adapted *inferential erotetic logic* (IEL). We utilize the framework of IEL and show some properties, relationships, and possible generalizations.

2.1.1 ADAPTED SET-OF-ANSWERS METHODOLOGY IN IEL

Inferential erotetic logic accepts only the first two of Hamblin's postulates and tries to keep the maximum of its (classical) declarative logic and its consequence relation. On the syntactic level of a considered formalized language, a question is assigned to a set of sentences (direct answers). Direct answers are declarative formulas and every question has at least two direct answers. Any finite and at least two-element set of sentences is the set of direct answers of some question [53, p. 11].

Let us apply our SAM introduced in the previous chapter. We define a (general) erotetic language $\mathcal{L}_Q$. A (general) declarative language $\mathcal{L}$ is extended by curly brackets ({, }) and question mark (?).[1] In the correspondence with Section 1.2, a question Q is the following structure

$$?\{\alpha_1, \alpha_2, ...\}$$

For the set of direct answers to question Q, we will use the symbol dQ. Let us recall that direct answers are syntactically distinct and $|dQ| \geq 2$. Moreover, we require here that the elements of dQ are declarative sentences.

In the case of finite versions of questions $?\{\alpha_1, ..., \alpha_n\}$ we suppose that the direct answers listed are semantically non-equivalent. The class of finite questions corresponds to the class of *questions of the first kind* in [51]. Some of them are important in our future examples and counterexamples. Let us mention two abbreviations and terms that will be very frequent in this book.

- *Simple yes-no questions* are of the form $?\alpha$, which is an abbreviation for $?\{\alpha, \neg\alpha\}$. If α is an atomic formula, then the term *atomic yes-no question* is used.
- A *conjunctive question* $?|\alpha, \beta|$ requires the answer whether α (and not β), or β (and not α), or neither α nor β, or both (α and β). It is an abbreviation for $?\{(\alpha\wedge\beta), (\neg\alpha\wedge\beta), (\alpha\wedge\neg\beta), (\neg\alpha\wedge\neg\beta)\}$. Similar versions are $?|\alpha, \beta, \gamma|$, $?|\alpha, \beta, \gamma, \delta|$, and so on.[2]

In the original version of IEL, questions are not identified with sets of direct answers: questions belong to an object-level language and are expressions of a strictly defined form, but the form is designed in such a way that, on the metalanguage level (and only here), the expression which occurs after the question mark designates the set of direct answers to the question. Questions are defined in such a way that sets of direct answers to them are explicitly specified. The general framework of IEL allows for other ways of formalizing questions.[3]

To avoid misunderstanding, we will use the following metavariables in this chapter:

- small Greek letters ($\alpha, \beta, \varphi, ...$) for declarative sentences,
- $Q, Q_1, ...$ for questions,
- capital Greek letters ($\Gamma, \Delta, ...$) for sets of declaratives, and
- $\Phi, \Phi_1, ...$ for sets of questions.

1 In this chapter we use only propositional examples in the language with the usual connectives ($\wedge, \vee, \rightarrow, \neg$).

2 Original IEL uses the symbol $?\pm|\alpha, \beta|$, etc. The concept of conjunctive questions was introduced in [45].

3 Personal communication with Andrzej Wiśniewski.

2.1.2 CONSEQUENCE RELATIONS IN IEL

Consequence relations are the central point of logic. Declarative logic can be defined by its consequence relation as a set of pairs $\langle \Gamma, \Delta \rangle$, where Γ and Δ are sets of (declarative) formulas and Δ is usually considered to be a singleton. Inferential erotetic logic makes one more step and adds new consequence relations mixing declaratives and interrogatives. The most important relations, which we are going to introduce, are the following: *evocation*, *erotetic implication*, and *reducibility*.

- *Evocation* is a binary relation $\langle \Gamma, Q \rangle$ between a set of declaratives Γ and a question.
- *Erotetic implication* is a ternary relation $\langle Q, \Gamma, Q_1 \rangle$ between an initial question Q and an implied question Q_1 with respect to a set of declaratives Γ.
- *Reducibility* is a ternary relation $\langle Q, \Gamma, \Phi \rangle$ between an initial question Q and a set of questions Φ with respect to a set of declaratives Γ.

Motivations and natural-language examples of these consequence relations will be introduced in the succeeding sections of this chapter. From the literature, let us recommend texts [51, 53] for both the evocation and erotetic implication; the concept of reducibility is studied in [23, 51, 57].

Our aim is to study erotetic consequence relations in a very general manner, independently of background logic. Nonetheless, the definitions of IEL consequence relations are based on the semantic entailment and the model-based approach, which must be understood as relative to a chosen logical background.

2.1.3 MODEL-BASED APPROACH

The following model-based approach was inspired by *minimal erotetic semantics* from [53]. Let us introduce the set of all models for a declarative language:

$$\mathcal{M}_{\mathsf{L}} = \{\mathbf{M} \mid \mathbf{M} \text{ is a (semantic) } \textit{model} \text{ for } \mathcal{L}\}.$$

The meaning of the term *model* varies dependently on the background logic L. If L is classical propositional logic (CPL, for short), then $\mathcal{M}_{CPL}$ is a set of all *valuations*. In the case of predicate logic, it is a set of all *structures* with a realization of non-logical symbols.

Because of the possibility of putting some additional constraints upon models (e.g., finiteness of models or preferences among models), we use a set $\mathcal{M} \subseteq \mathcal{M}_{\mathsf{L}}$ as a 'general' case. If necessary, the background logic and restrictions posed on models will be stated explicitly.

Speaking about the *tautologies* of logic L, we mean the set of formulas

$$\textsc{Taut}_{\mathsf{L}} = \{\varphi \mid (\forall \mathbf{M} \in \mathcal{M}_{\mathsf{L}})(\mathbf{M} \models \varphi)\}.$$

If a restricted set of models $\mathcal{M}$ is in use, we speak about $\mathcal{M}$*-tautologies*

$$\text{TAUT}_{\mathsf{L}}^{\mathcal{M}} = \{\varphi \mid (\forall \mathbf{M} \in \mathcal{M})(\mathbf{M} \vDash \varphi)\}.$$

All semantic terms may be relativized to $\mathcal{M}$. Each declarative sentence φ (in the language $\mathcal{L}$) has its (restricted) set of models

$$\mathcal{M}^{\varphi} = \{\mathbf{M} \in \mathcal{M} \mid \mathbf{M} \vDash \varphi\}$$

and similarly for a set of sentences Γ

$$\mathcal{M}^{\Gamma} = \{\mathbf{M} \in \mathcal{M} \mid (\forall \gamma \in \Gamma)(\mathbf{M} \vDash \gamma)\}.$$

Semantic entailment

Let us recall the (semantic) entailment relation ($\vDash$). For any set of formulas Γ and any formula ψ:

$$\Gamma \vDash \psi \text{ iff } \mathcal{M}^{\Gamma} \subseteq \mathcal{M}^{\psi}.$$

In case of $\Gamma = \{\varphi\}$, we write only $\varphi \vDash \psi$, which means $\mathcal{M}^{\varphi} \subseteq \mathcal{M}^{\psi}$.

Multiple-conclusion entailment relation ($\Vdash$, mc-entailment, for short) is defined in a similar way:

$$\Gamma \Vdash \Delta \text{ iff } \mathcal{M}^{\Gamma} \subseteq \bigcup_{\delta \in \Delta} \mathcal{M}^{\delta}$$

If $\mathcal{M}^{\Gamma} = \mathcal{M}^{\Delta}$, let us write $\Gamma \equiv \Delta$.[4]

Mc-entailment is reflexive ($\Gamma \Vdash \Gamma$), but it is neither a symmetric nor a transitive relation.

Example 1. *Let $\Gamma \subseteq \text{TAUT}_{\mathsf{L}}^{\mathcal{M}}$, Δ be a set of sentences containing at least one tautology and at least one contradiction, and Σ be such that $\bigcup_{\sigma \in \Sigma} \mathcal{M}^{\sigma} \subset \mathcal{M}$. Then $\Gamma \Vdash \Delta$ and $\Delta \Vdash \Sigma$, but $\Gamma \nVdash \Sigma$.*

Entailment is definable by mc-entailment:

$$\Gamma \vDash \varphi \text{ iff } \Gamma \Vdash \{\varphi\}$$

However, mc-entailment is not definable by entailment. In this context, the following proposition could be a surprise at first sight:

Proposition 1. *Entailment (for logic* L*) is compact iff mc-entailment (for* L*) is compact.*

Proof. See [51, pp. 109–110].[5] □

---

4 In the case of the semantic equivalence of formulas φ and ψ, we write $\varphi \equiv \psi$. Let us remark that two different sets of models do not imply the existence of two different sets of sentences (in $\mathcal{L}$).

5 We say that mc-entailment is compact iff for each $\Gamma \Vdash \Delta$ there are finite subsets $\Gamma' \subseteq \Gamma$ and $\Delta' \subseteq \Delta$ such that $\Gamma' \Vdash \Delta'$.

2.1.4 BASIC PROPERTIES OF QUESTIONS

After we have introduced the SAM representation of questions and the model-based approach, we can mention some basic properties of questions. First, let us introduce the term *soundness*, which is one of the most important terms in IEL.

Definition 1. *A question Q is* sound in **M** *iff* $\exists \alpha \in dQ$ *such that* $\mathbf{M} \models \alpha$.

A question is sound with respect to model **M** whenever it has at least one direct answer true in **M**. See [51, p. 113].

For all IEL consequence relations, it is important to state the soundness of a question with respect to the set of declaratives.

Definition 2. *A question Q is* sound relative to Γ *iff* $\Gamma \Vdash dQ$.

The sum of all classes of models of each direct answer α, i.e., $\bigcup_{\alpha \in dQ} \mathcal{M}^{\alpha}$, is called the *semantic range* of question Q. Our liberal approach admits some 'curious' questions; one of them is a *completely contradictory question* that has only contradictions in its set of direct answers, its semantic range is just an empty set. Another type is a question with the some tautology among its direct answers, then the semantic range expands to the whole $\mathcal{M}$. Questions with such a range are called *safe*.[6] Of course, it need not be any tautology among direct answers for it to be a safe question.

Definition 3.

- *A question Q is* safe *iff* $\bigcup_{\alpha \in dQ} \mathcal{M}^{\alpha} = \mathcal{M}$.
- *A question Q is* risky *iff* $\bigcup_{\alpha \in dQ} \mathcal{M}^{\alpha} \subset \mathcal{M}$.

Questions $?\alpha, ?|\alpha, \beta|$ are safe in CPL, but neither is safe in Bochvar logic. If β is not equivalent to $\neg\alpha$, then $?\{\alpha, \beta\}$ is risky in CPL. Neither $?\alpha$ nor $?|\alpha, \beta|$ are safe in intuitionistic logic, but there are safe questions in this logic (e.g., each question with at least one tautology among direct answers). Simple yes-no questions are safe in logics that accept the law of an excluded middle.

It is good to emphasize that the set of direct answers of a safe question is mc-entailed by every set of declaratives. On the other hand, knowing a question to be sound relative to every set of declaratives implies its safeness.

Fact 1. *Q is safe iff* $(\forall\Gamma)(\Gamma \Vdash dQ)$.

Safe questions are sound relative to $\Gamma = \emptyset$.

Fact 2. *If* $\emptyset \models Q$, *then Q is safe.*

---

6 This term originates from Nuel Belnap.

2.1.5 THE ROAD WE ARE GOING TO TAKE

After the introducing *evocation* and *presupposition* of a question in Section 2.2, we will show the role of maximal and prospective presuppositions in relationship to the semantic range of questions. Some classes of questions will be based on the concept of presupposition.

Section 2.3 is crucial from the viewpoint of inferences in IEL. We investigate *erotetic implication* and *reducibility* there. An important part is devoted to the discussion of the role of an auxiliary set of declaratives. We will demonstrate some variants of erotetic implication and their properties. The chosen formal shape of questions in IEL makes it possible to compare questions in the sense of their *answerhood power*. Inspired by [16] and [51, section 5.2.3] we will examine the relationship of 'giving an answer' to one question to another, which is a generalisation of Kubiński's term 'weaker question'.

Questions will be considered as independent structures not being combined by logical connectives. Perhaps, reducibility could substitute a form of connection of questions. This brings us to the last note on the use of the symbols $\models$ and $\vdash$. Because of the clear border between the declarative and interrogative parts of the language $\mathcal{L}_Q$, we will use them in many meanings. However, the meaning will be transparent by the context of the visual symbols.

2.2 QUESTIONS AND DECLARATIVES

In this section, we introduce two terms: *evocation* and *presupposition*. The first one will provide a consequence relation between a set of declaratives and a question. The second one is an important term in almost all logics of questions and some classes of questions are based on it.

2.2.1 EVOCATION

After a lecture, we expect a lecturer to be ready to answer some questions that were 'evoked' by his or her talk. *Evocation* seems to be one of the most obvious relationships among declarative sentences and questions. Of course, next to the connection is: *question—answer*. Almost every piece of information can give rise to a question. What is the aim of such a question?

First, it should complete our knowledge in some direction. Asking a question, we want to get more than by the conclusion based on background knowledge. A question Q should be *informative* relative to a set of declaratives, this means there is no direct answer to Q, which is a consequence of this set of declaratives.

Second, after answering an evoked question the answer must be consistent within the evoking context. Moreover, *transmission of truth into soundness* is required: if an evoking set of declaratives has a model, there must be at least one direct answer of the evoked question that is true in this model. An evoked question should be sound relative to an evoking set of declaratives.[7]

The definition of *evocation* is based on the previous two points (cf. [51, 53]). A question Q is evoked by a set of declaratives Γ if Q is sound and informative relative to Γ:

Definition 4 (Evocation). *A set of declarative sentences* Γ evokes *a question* Q *(let us write* $\Gamma \models Q$*) iff*

1. $\Gamma \Vdash\!\models dQ$,
2. $(\forall \alpha \in dQ)(\Gamma \not\models \alpha)$.

In our model-based approach we can rewrite both conditions this way:

1. $\mathcal{M}^{\Gamma} \subseteq \bigcup_{\alpha \in dQ} \mathcal{M}^{\alpha}$
2. $(\forall \alpha \in dQ)(\mathcal{M}^{\Gamma} \nsubseteq \mathcal{M}^{\alpha})$

In some special cases (e.g., dQ is finite or entailment is compact) we can define evocation without the link to mc-entailment. The first condition is of the form: there are $\alpha_1, \ldots, \alpha_n \in dQ$ such that $\Gamma \models \bigvee_1^n \alpha_i$.

This is the case of one of our introductory examples. Let us remind the group of three card players. The context

> Γ: *Either Anne has the Joker or Bill has the Joker or Catherine has the Joker.*

evokes the question

> Q: *Who has the Joker: Anne, Bill, or Catherine?*

with direct answers

> α: *Anne has the Joker.*
> β: *Bill has the Joker.*
> γ: *Catherine has the Joker.*

In particular, Γ can consist of one formula, disjunction of direct answers of Q, i.e., $(\alpha \vee \beta \vee \gamma)$. The first condition of evocation is clearly satisfied and the second one is satisfied because no direct answer is entailed by Γ.

Evocation provides some useful properties of both an evoked question and a set of declaratives. The following fact lists some of them:

---

7 In this chapter, epistemic issues are not important for us, so, we shall not use the word 'knowledge' for a declarative background. We will use phrases like 'set of declarative(s) (sentences)' or 'database' instead.

Fact 3. *If* $\Gamma \models Q$, *then*

- Γ *is not a contradictory set,*
- *there is no tautology in* dQ, *and*
- Q *is not a completely contradictory question.*[8]

Thanks to Fact 1, we obtain a not very intuitive consequence: every safe question is evoked by any set of declaratives, which does not entail any direct answer of the question. It punctuates the special position of safe questions and their semantic range. If we restrict the definition of evocation to just risky questions, we obtain the definition of *generation.*[9]

Generation does not solve all problems with irrelevant and inefficient evoked questions either. We can accept another restriction to avoid questions that have direct answers which are incompatible with the declaratives in Γ. Borrowing an example from [7], the set of declaratives$\{\alpha \vee \beta, \gamma\}$ also evokes the question $?\{\alpha, \beta, \neg\gamma\}$. To eliminate this, the consistency of each direct answer with respect to Γ would be required, i.e., we could add one more condition to Definition 4:

$$(\forall \alpha \in dQ)(\mathcal{M}^{\Gamma} \cap \mathcal{M}^{\alpha} \neq \emptyset)$$

Some solutions of the problem of irrelevant, and inefficient questions based on semantics in the background are discussed in the just mentioned paper [7]. For our purpose, the study of consequence relations in IEL, we will keep Definition 4 unchanged.

As a conclusion of the semantic definition of evocation, we have the following expected behavior of evocation: semantically equivalent databases evoke the same questions.

Fact 4. *For every* Γ, Δ *and* Q, *if* $\Gamma \equiv \Delta$, *then* $\Gamma \models Q$ *iff* $\Delta \models Q$.

If Γ evokes Q, then we have to be careful in concluding that there is a subset $\Delta \subseteq \Gamma$ such that Δ evokes Q, see the first item in the following fact.

Fact 5. *If* $\Gamma \models Q$ *and* $\Delta \subseteq \Gamma \subseteq \Sigma$, *then*

- $\Delta \models Q$ *if* $\Delta \Vdash dQ$,
- $\Sigma \models Q$ *if* $(\forall \alpha \in dQ)(\Sigma \not\models \alpha)$.

The second item points out the non-monotonicity of evocation (in declaratives). Considering questions as sets of answers, evocation is non-monotonic in interrogatives as well, see Section 2.3.3.

Fact 6. *If* $\Gamma \models Q$ *and the entailment is compact, then* $\Delta \models Q_1$ *for some finite subset* dQ_1 *of* dQ *and some finite subset* Δ *of* Γ.

8 Completely contradictory questions have only contradictions as direct answers.

9 Properties of *generation* are discussed in [51, chapter 6].

These and some more properties of evocation (and generation) are discussed in the book [51].

2.2.2 PRESUPPOSITIONS

Many properties of questions are based on the concept of *presupposition*. Everyone who has attended a basic course in research methods in the social sciences has heard of the importance to consider presuppositions of questions in questionaries.

If we are asked *Who has the Joker: Anne, Bill, or Catherine?*, we can recognize that it is presupposed that Anne has it or Bill has it or Catherine has it. What is presupposed must be valid under each answer to a question. Moreover, an answer to a question should bring at least the same information as the presupposition does. The following definition (originally given by Nuel Belnap) is from [51]:

Definition 5. *A declarative formula φ is a* presupposition *of a question Q iff* $(\forall\alpha \in dQ)(\alpha \vDash \varphi)$.

A presupposition of a question is entailed by each direct answer to the question. Let us write Pres$Q$ for the set of all presuppositions of Q.

At the first sight, the set Pres$Q$ could contain a lot of sentences. Let us have a question $Q = ?\{\alpha_1, \alpha_2\}$, the set of presuppositions (e.g., in CPL) contains $(\alpha_1 \vee \alpha_2)$, $(\alpha_1 \vee \alpha_2 \vee \varphi)$, $(\alpha_1 \vee \alpha_2 \vee \neg\varphi)$, etc. Looking at the very relevant member $(\alpha_1 \vee \alpha_2)$ it is useful to introduce the concept of *maximal presupposition*. Formula $(\alpha_1 \vee \alpha_2)$ entails each presupposition of the question Q.

Definition 6. *A declarative formula φ is a* maximal presupposition *of a question Q iff $\varphi \in \mathsf{Pres}Q$ and $(\forall\psi \in \mathsf{Pres}Q)(\varphi \vDash \psi)$.*

The model-theoretical view shows it in a direct way. The definition of presupposition gives $\bigcup_{\alpha\in dQ} \mathcal{M}^{\alpha} \subseteq \mathcal{M}^{\varphi}$, for each $\varphi \in \mathsf{Pres}Q$, which means

$$\bigcup_{\alpha\in dQ} \mathcal{M}^{\alpha} \subseteq \bigcap_{\varphi\in \mathsf{Pres}Q} \mathcal{M}^{\varphi} = \mathcal{M}^{\mathsf{Pres}Q}$$

and the set $\mathcal{M}^{\mathsf{Pres}Q}$ is a model-based counterpart to the definition of maximal presuppositions.

If the background logic has tautologies, each of them is in Pres$Q$.

$$\textsc{Taut}_{\mathsf{L}}^{\mathcal{M}} \subseteq \mathsf{Pres}Q$$

Considering safe questions we obtain

Fact 7. *If Q is safe, then* $\mathsf{Pres}Q = \textit{TAUT}^{\mathcal{M}}_{\mathsf{L}}$.

This fact says that if Q is safe, then $\bigcup_{\alpha \in dQ} \mathcal{M}^{\alpha} = \mathcal{M}^{\mathsf{Pres}Q}$. In classical propositional logic the disjunction of all direct answers of a question is a presupposition of this question and if $\mathsf{Pres}Q = \mathrm{TAUT}^{\mathcal{M}}_{\mathrm{CPL}}$, then Q is safe. This evokes a (meta)question of whether the implication from right to left is valid. If Q is not safe, then we know that $\bigcup_{\alpha \in dQ} \mathcal{M}^{\alpha}$ is a proper subset of $\mathcal{M}$. But what about $\mathcal{M}^{\mathsf{Pres}Q}$? After introducing a class of *normal questions* (see page 31) $\mathcal{M}^{\mathsf{Pres}Q} \subset \mathcal{M}$ will be valid, as well as the implication from right to left (see Fact 9).

A presupposition can be seen as information, which is announced by asking a question without answering it. Such information is relatively small. The semantic range of all maximal presuppositions is wider than the range of a question. Looking at finite CPL example where the disjunction of all direct answers forms just the semantic range of the question brings us to the idea of *prospective presupposition*. It is a presupposition which question Q is sound relative to.

Definition 7. *A declarative formula φ is a* prospective presupposition *of a question Q iff $\varphi \in \mathsf{Pres}Q$ and $\varphi \Vdash dQ$. Let us write $\varphi \in \mathsf{PPres}Q$.*

All prospective presuppositions of a question are equivalent.

Proposition 2. *If $\varphi, \psi \in \mathsf{PPres}Q$, then $\varphi \equiv \psi$.*

Proof. Let us prove that $\varphi \vDash \psi$. If $\mathbf{M} \vDash \varphi$, then there is $\alpha \in dQ$ such that $\mathbf{M} \vDash \alpha$. Since $\psi \in \mathsf{Pres}Q$, $\alpha \vDash \psi$, and it gives $\mathbf{M} \vDash \psi$, we get $\varphi \vDash \psi$.

$\psi \vDash \varphi$ is proved by the same way. □

A prospective presupposition forms exactly the semantic range of a question:

$$\bigcup_{\alpha \in dQ} \mathcal{M}^{\alpha} = \mathcal{M}^{\mathsf{PPres}Q}$$

If Q has a prospective presupposition, it can be understood as the 'informatively strongest' presupposition.

Two questions with the same sets of presuppositions have the same prospective presuppositions.

Proposition 3. *If $\mathsf{Pres}Q = \mathsf{Pres}Q_1$ and both $\mathsf{PPres}Q$ and $\mathsf{PPres}Q_1$ are not empty, then $\mathsf{PPres}Q = \mathsf{PPres}Q_1$.*

Proof. We show that if $\varphi \in \mathsf{PPres}Q$ and $\psi \in \mathsf{PPres}Q_1$, then $\varphi \equiv \psi$.

$\varphi \in \mathsf{PPres}Q$ implies $\mathcal{M}^{\varphi} = \bigcup_{\alpha \in dQ} \mathcal{M}^{\alpha}$ and $\bigcup_{\alpha \in dQ} \mathcal{M}^{\alpha} \subseteq \mathcal{M}^{\psi}$, because $\psi \in \mathsf{Pres}Q$. It gives $\mathcal{M}^{\varphi} \subseteq \mathcal{M}^{\psi}$ and $\varphi \vDash \psi$.

The proof that $\psi \vDash \varphi$ is similar. □

Presuppositions of evoked questions are entailed by the evoking set of declaratives.

Fact 8. *If* $\Gamma \models Q$, *then* $\Gamma \models \varphi$, *for each* $\varphi \in \mathsf{Pres}Q$.

The implication from right to left does not hold. If we only know $\mathcal{M}^{\Gamma} \subseteq \mathcal{M}^{\mathsf{Pres}Q}$, we are not sure about $\mathcal{M}^{\Gamma} \subseteq \bigcup_{\alpha \in dQ} \mathcal{M}^{\alpha}$ as required by the first condition of evocation. Clearly, the informativeness must be ensured as well. Let us note that it cannot be improved by replacing Pres$Q$ with PPres$Q$. We will return to this in the next subsection studying the topic of normal questions. To sum up all properties in interplay of a question evoked by Γ, its semantic range and presuppositions give us a look at the diagram:

$$\mathcal{M}^{\Gamma} \subseteq \bigcup_{\alpha \in dQ} \mathcal{M}^{\alpha} = \mathcal{M}^{\mathsf{PPres}Q} \subseteq \mathcal{M}^{\mathsf{Pres}Q}$$

Classes of questions based on presuppositions

Using the term *presupposition* we can define some classes of questions. Names and definitions of the following classes are from [51]. We only add the model-based approach and make transparent some results on presuppositions and evocations, which are presented in chapters 4 and 5 of the book [51].

Normal questions A question Q is called *normal* if it is sound relative to its set of presuppositions, $\mathsf{Pres}Q \Vdash dQ$. A model-based approach introduces normal questions as questions with a semantic range delimitated by models of maximal presuppositions.

- $Q \in$ NORMAL iff $\bigcup_{\alpha \in dQ} \mathcal{M}^{\alpha} = \mathcal{M}^{\mathsf{Pres}Q}$

Working with finite sets of direct answers and in logical systems with the 'classical behavior' of disjunction (each direct answer entails the disjunction of all direct answers) we do not leave the class NORMAL. Non-normal questions can be found in classical predicate logic.

The class of safe questions is a subset of the class of normal questions.

SAFE $\subseteq$ NORMAL

Two additional facts follow to shut away the discussions we have started with Facts 7 and 8.

Fact 9. *If* $\mathsf{Pres}Q = \textit{Taut}^{\mathcal{M}}_{\mathsf{L}}$ *and* Q *is normal, then* Q *is safe.*

Fact 10. *If* $\Gamma \models \varphi$, *for each* $\varphi \in \mathsf{Pres}Q$, *and* $\Gamma \not\models \alpha$, *for each* $\alpha \in dQ$ *of a normal question* Q, *then* $\Gamma \models Q$.

Fact 10, together with Fact 8, give the conditions for evocation of normal questions.[10]

10 Cf. Theorem 5.23 in [51].

Regular questions Each question with the non-empty set of prospective presuppositions is *regular.*

- $Q \in$ REGULAR iff $(\exists\varphi \in \mathsf{Pres}Q)(\varphi \Vdash dQ)$

The regularity of Q gives $\mathcal{M}^{\mathsf{Pres}Q} \subseteq \mathcal{M}^{\varphi} \subseteq \bigcup_{\alpha\in dQ} \mathcal{M}^{\alpha}$ and it holds

$$\text{REGULAR} \subseteq \text{NORMAL}$$

If entailment is compact, both classes are equal.

Normal questions are sound relative to $\mathsf{Pres}Q$ and regular questions are sound relative to $\mathsf{PPres}Q$. The following example shows an expected fact that regularity of a question is still not sufficient to evoce the question from its set of presuppositions.

Example 2 (in CPL). *Let* $Q = ?\{(\alpha \vee \beta), \alpha\}$. *This question is normal and regular, the formula* $(\alpha \vee \beta)$ *is a prospective presupposition of* Q*, but* $\mathsf{Pres}Q \nvDash Q$.

If there is a set of declaratives Γ such that $\Gamma \vDash Q$, then normal (regular) questions are sound as well as informative relative to $\mathsf{Pres}Q$ ($\mathsf{PPres}Q$). This is summed up by

Proposition 4. *Let* $\Gamma \vDash Q$*, for some set of declaratives* Γ. *Then*

1. $Q \in$ *NORMAL implies* $\mathsf{Pres}Q \vDash Q$.
2. $Q \in$ *REGULAR implies* $\varphi \vDash Q$*, for* $\varphi \in \mathsf{PPres}Q$.

Proof. For the first item, only informativeness (relative to $\mathsf{Pres}Q$) must be showed. But if it is not valid, then Fact 8 causes non-informativeness of Q relative to Γ.

The second item is proved by the same idea. □

Self-rhetorical questions Another special class of questions are *self-rhetorical* ones. They have at least one direct answer entailed by the set of presuppositions.

- $Q \in$ SELF-RHETORICAL iff $(\exists\alpha \in dQ)(\mathsf{Pres}Q \vDash \alpha)$

It is clear that self-rhetorical questions are normal. However, do we ask such questions? This class includes such strange questions as *completely contradictory* questions, which have only contradictions in the set of direct answers, and questions with tautologies among direct answers.

An evoked question is not of this kind.

Proposition 5. *If there is* Γ *such that* $\Gamma \vDash Q$*, then* Q *is not self-rhetorical.*

Proof. From Fact 8. □

Proper questions Normal and not self-rhetorical questions are called *proper*. Proper questions are evoked by their set of presuppositions.

- $Q \in$ PROPER iff Pres$Q \models Q$

Evoked normal questions are proper (compare both Proposition 4 and Proposition 5) and this makes the PROPER class prominent. The set of all presuppositions of a question is believed to be a natural (declarative) context for evocation of the question.

2.3 QUESTIONS AND QUESTIONS

This section is devoted to (1) inferential structures, in which questions appear on both sides, i.e., among premises as well as conclusions (*erotetic implication* and *reducibility of questions to sets of questions* will be introduced); and (2) to relations between two questions based on their sets of direct answers. The second point focuses on (the still vague) term 'answerhood power' of questions formalized by the adapted set-of-answers methodology.

2.3.1 EROTETIC IMPLICATION

Now, we extend the class of inferences by 'implication' between two questions with a possible assistance of a set of declaratives. Let us start with an easy though a bit tricky example. If I ask

> Q: *What is Peter a graduate of: a faculty of law or a faculty of economics?*

then I can be satisfied by the answer

> *He is a lawyer.*

even if I did not ask

> Q_1: *What is Peter: a lawyer or an economist?*

The connection between both questions could be stated by the following set of declaratives:

> *Someone is a graduate of a faculty of law iff he/she is a lawyer.*
> *Someone is a graduate of a faculty of economics iff he/she is an economist.*

The first question Q can be formalized by $?\{\alpha_1, \alpha_2\}$ and the latter one, speaking of Peter's position, can be $?\{\beta_1, \beta_2\}$. Looking at the questions there is no connection between them. The relationship is based on the set of declaratives $\Gamma = \{(\alpha_1 \leftrightarrow \beta_1), (\alpha_2 \leftrightarrow \beta_2)\}$. The question Q *implies* question Q_1 on the basis of Γ and such relationship will be written as

$$\Gamma, Q \vDash Q_1$$

The introduced relation is called *erotetic implication* (*e-implication*, for short). It is defined by the validity of two clauses:[11]

Definition 8 (E-implication). *A question Q implies a question Q_1 on the basis of a set of declaratives Γ iff*

1. $(\forall\alpha \in dQ)(\Gamma \cup \alpha \Vdash dQ_1)$,
2. $(\forall\beta \in dQ_1)(\exists\Delta \subset dQ)(\Delta \neq \emptyset$ *and* $\Gamma \cup \beta \Vdash \Delta)$.

The first clause expresses the soundness of an implied question relative to each extension of Γ by $\alpha \in dQ$. This *transmission of truth/soundness into soundness* has the following meaning: if there is a model of Γ and a direct answer to Q, then there must be a direct answer to Q_1 that is valid in this model. If Q_1 is safe, then this condition is always valid (see Fact 1).

The second clause requires direct answers to Q_1 to be *cognitively useful* in restricting the set of direct answers of the implying question Q.

In our introductory example, both questions are *erotetically equivalent* with respect to Γ, i.e., $\Gamma, Q \vDash Q_1$ as well as $\Gamma, Q_1 \vDash Q$.

In comparison with evocation, the role of the set of declaratives is a bit different, it plays, especially, an auxiliary role. E-implication is monotonic in declaratives [51, p. 173]:

Fact 11. *If $\Gamma, Q \vDash Q_1$, then $\Delta, \Gamma, Q \vDash Q_1$, for any set of declaratives Δ.*

This fact could be called *weakening in declaratives.*[12]

It deserves to say more about the role of auxiliary sets of declaratives. We will do this in the following subsection.

Pure erotetic implication

Pure e-implication is erotetic implication with an empty set of declaratives. From Fact 11, whenever two questions are in a relation of pure e-implication, then they are in the relation of e-implication for any set of declaratives.

If one question purely e-implies another question, then both questions have the same semantic range:

Proposition 6. *If $Q \vDash Q_1$, then $\bigcup_{\alpha \in dQ} \mathcal{M}^\alpha = \bigcup_{\beta \in dQ_1} \mathcal{M}^\beta$.*

---

11 The definition is from [51]. We will write shortly: $\Gamma \cup \varphi$ instead of $\Gamma \cup \{\varphi\}$.

12 For any questions Q and Q_1 and a contradictory set $\bot$ it holds that $\bot, Q \vDash Q_1$.

Proof. From the first condition of Definition 8 we obtain

$$\bigcup_{\alpha \in dQ} \mathcal{M}^{\alpha} \subseteq \bigcup_{\beta \in dQ_1} \mathcal{M}^{\beta}$$

From the second one

$$\bigcup_{\beta \in dQ_1} \mathcal{M}^{\beta} \subseteq \bigcup_{\Delta} \bigcup_{\alpha \in \Delta} \mathcal{M}^{\alpha},$$

for each non-empty $\Delta \subset dQ$ relevant to βs. Moreover, $\bigcup_{\alpha \in dQ} \mathcal{M}^{\alpha}$ includes $\bigcup_{\Delta} \bigcup_{\alpha \in \Delta} \mathcal{M}^{\alpha}$ not only for relevant Δs, thus,

$$\bigcup_{\beta \in dQ_1} \mathcal{M}^{\beta} \subseteq \bigcup_{\Delta} \bigcup_{\alpha \in \Delta} \mathcal{M}^{\alpha} \subseteq \bigcup_{\alpha \in dQ} \mathcal{M}^{\alpha}.$$

□

From this proposition we can conclude that classes of safe and risky questions are closed under pure e-implication for both implied and implying questions:[13]

Fact 12. *If $Q \vDash Q_1$, then Q is safe (risky) iff Q_1 is safe (risky).*

The same semantic range of questions linked together by pure e-implication does not form an equivalence relation on questions (see non-symmetry in Example 4 and non-transitivity in Example 5, page 38). Nonetheless, pure e-implication has some important consequences for classes of presuppositions.[14]

Proposition 7. *If $Q \vDash Q_1$, then* $\mathrm{Pres}Q = \mathrm{Pres}Q_1$.

Proof. First, let us prove $\mathrm{Pres}Q \subseteq \mathrm{Pres}Q_1$. Suppose $\varphi \in \mathrm{Pres}Q$, so $\bigcup_{\alpha \in dQ} \mathcal{M}^{\alpha} \subseteq \mathcal{M}^{\varphi}$. Simultaneously, we know that from the second condition of the definition of pure e-implication there is a non-empty $\Delta \subset dQ$, for each $\beta \in dQ_1$, such that $\mathcal{M}^{\beta} \subseteq \bigcup_{\alpha \in \Delta} \mathcal{M}^{\alpha} \subseteq \bigcup_{\alpha \in dQ} \mathcal{M}^{\alpha}$. Thus, $\mathcal{M}^{\beta} \subseteq \mathcal{M}^{\varphi}$, for each $\beta \in dQ_1$.

Second, for proving $\mathrm{Pres}Q_1 \subseteq \mathrm{Pres}Q$ suppose $\varphi \in \mathrm{Pres}Q_1$. The following inclusions are valid $\mathcal{M}^{\alpha} \subseteq \bigcup_{\beta \in dQ_1} \mathcal{M}^{\beta} \subseteq \mathcal{M}^{\varphi}$, for each $\alpha \in dQ$. □

The claim of Proposition 7 is not valid for general e-implication (cf. undermentioned Example 3). On this proposition we can base the following statements about classes of normal and regular questions.

Proposition 8. *If $Q \vDash Q_1$, then Q is normal iff Q_1 is normal.*

13 Cf. Theorem 7.29 in [51, p. 184]

14 See the same result in [51, p. 184], Theorem 7.33.

Proof. If Q is normal, then

$$\bigcup_{\beta \in dQ_1} \mathcal{M}^{\beta} = \bigcup_{\alpha \in dQ} \mathcal{M}^{\alpha} = \mathcal{M}^{\mathsf{Pres}Q} = \mathcal{M}^{\mathsf{Pres}Q_1}$$

The first equation is from Proposition 6, the second one is from the normality of Q, and the third one is from Proposition 7. □

Proposition 9. *If $Q \vDash Q_1$, then Q is regular iff Q_1 is regular.*

Proof. Let us suppose $Q \vDash Q_1$ and Q is regular. From $Q \vDash Q_1$ and Proposition 6 we obtain $\bigcup_{\alpha \in dQ} \mathcal{M}^{\alpha} = \bigcup_{\beta \in dQ_1} \mathcal{M}^{\beta}$, which means that $\mathcal{M}^{\mathsf{PPres}Q} = \mathcal{M}^{\mathsf{PPres}Q_1}$. The regularity of Q implies that there is a formula $\varphi \in \mathsf{PPres}Q$. Putting it together, $\mathcal{M}^{\varphi} = \bigcup_{\beta \in dQ_1} \mathcal{M}^{\beta}$. Thus, $\varphi \in \mathsf{PPres}Q_1$. □

We obtained a similar result for safe and risky questions in Fact 12. Classes of normal and regular questions are closed with respect to pure e-implication. Normal (regular) questions purely imply only normal (regular) questions and they are purely implied by the same kind of questions.[15]

As a final remark, let us add that whenever $Q \vDash Q_1$, then Q is a completely contradictory question if and only if Q_1 is so.

Note on auxiliary sets of declaratives in e-implication Let us remember the introductory example on page 33 to recall the importance of declaratives for e-implication. Similarly, it can be crucial to display presuppositions explicitly. The following example will point out the role of implicitly and explicitly expressed presuppositions:

Example 3 (in CPL). *If $Q_1 = ?\{\alpha, \beta, \gamma\}$ and $Q_2 = ?\{\alpha, \beta\}$ (for atomic α, β, γ), then neither $Q_1 \vDash Q_2$ nor $Q_2 \vDash Q_1$ (see the different semantic ranges of both questions). On the other hand, if we knew that it must be $(\alpha \vee \beta)$, then $(\alpha \vee \beta), Q_1 \vDash Q_2$.*

In this example, the question Q_2 is normal as well as regular, then $\mathsf{PPres}Q_2 \Vdash dQ_2$ and, in addition, there is Δ, a non-empty proper subset of dQ_1, such that $\mathsf{PPres}Q_2 \Vdash \Delta$. It gives $\mathsf{PPres}Q_2, Q_1 \vDash Q_2$. If the set $\mathsf{PPres}Q_2$ is *explicitly* expressed, the implication from Q_1 to Q_2 is justified.

But now, back to the general approach. In the following fact we display when we can say that two questions and a set of declaratives are in a relationship of e-implication.

Fact 13. *Let us have Γ and two questions Q_1 and Q_2. In order to conclude $\Gamma, Q_1 \vDash Q_2$ it is sufficient to have $\Gamma \Vdash dQ_2$ and $\Gamma \Vdash \Delta$, where Δ is a non-empty proper subset of dQ_1.*

---

15 Propositions 8 and 9 put together results included in [51, pp. 185–186].

This fact can be reread in the form:

> If Q_2 is sound relative to Γ and Γ gives a partial answer to Q_1, then Q_1 implies Q_2 with respect to Γ.

We will add some more points to this discussion in Section 2.3.2 and in the last paragraph of Section 2.3.4.

Regular erotetic implication

A special kind of erotetic implication arises if there is exactly one direct answer in each Δ in the second clause of e-impliction (Definition 8). Then we say that Q *regularly* implies Q_1 (on the basis of Γ). The following definition originates from [53, p. 26]:

Definition 9. *A question Q* regularly implies *a question Q_1 on the basis of a set of declaratives Γ iff*

1. $(\forall \alpha \in dQ)(\Gamma \cup \alpha \Vdash dQ_1)$,
2. $(\forall \beta \in dQ_1)(\exists \alpha \in dQ)(\Gamma \cup \beta \models \alpha)$.

Because of the special importance of this relation let us use the symbol $\vdash$ for it and write $\Gamma, Q \vdash Q_1$.

In the case of *pure regular e-implication*, both conditions are changed into the form:

1. $(\forall \alpha \in dQ)(\alpha \Vdash dQ_1)$,
2. $(\forall \beta \in dQ_1)(\exists \alpha \in dQ)(\beta \models \alpha)$.

If $Q \vdash Q_1$ and we can answer Q_1, then we have an answer to Q. The relationship of pure regular e-implication between two questions says that the implied question is 'stronger' than the implying one in the sense of answerhood (see also Section 2.3.3).

Regularity can be enforced by the minimal number of direct answers of an implying question: if $Q \models Q_1$ and $|dQ| = 2$, then $Q \vdash Q_1$.

Basic properties of erotetic implication

In this subsection, we are going to speak about the reflexivity, symmetry, and transitivity of e-implication.

Erotetic implication is a reflexive relation.

Fact 14. $\Gamma, Q \models Q$, *for each* Γ *and* Q.

Even if there are examples of the symmetric behavior of e-implication, it is not symmetric generally.

Example 4 (in CPL). *Let* $Q_1 = ?\{(\alpha \vee \beta), \alpha\}$ *and* $Q_2 = ?\{\alpha, \beta\}$*. Then* $Q_1 \vDash Q_2$*, but* $Q_2 \nvDash Q_1$*.*

In the example, there is no non-empty proper subset of dQ_2 for the formula $(\alpha \vee \beta)$ to fulfil the second condition in the definition of e-implication. Moreover, it is useful to add that Q_1 regularly implies Q_2.

Erotetic implication is not transitive either.

Example 5 (in CPL). $?(\alpha \wedge \beta) \vdash ?|\alpha, \beta|$ *and* $?|\alpha, \beta| \vDash ?\alpha$*, but* $?(\alpha \wedge \beta) \nvDash ?\alpha$*.*

On the other hand, if we consider regular e-implication only, the following proposition is valid.

Proposition 10. *If* $Q_1 \vdash Q_2$ *and* $Q_2 \vdash Q_3$*, then* $Q_1 \vdash Q_3$*.*

Proof. The first condition of Definition 9 is proved by Proposition 6.

The second clause of this definition is based on regularity that gives $(\forall \gamma \in dQ_3)(\exists \alpha \in dQ_1)(\mathcal{M}^\gamma \subseteq \mathcal{M}^\alpha)$. □

We can do a cautious strengthening with the following fact:

Fact 15. *If* $\Gamma, Q_1 \vdash Q_2$ *and* $Q_2 \vdash Q_3$*, then* $\Gamma, Q_1 \vdash Q_3$*.*

As a final remark, let us add that presuppositions of an implied question are entailed by each direct answer of an implying question (with respect to an auxiliary set of declaratives).

Fact 16. *Let* $\Gamma, Q \vDash Q_1$*. Then*

1. $(\forall \alpha \in dQ)(\forall \varphi \in \mathrm{Pres}Q_1)(\Gamma \cup \alpha \vDash \varphi)$
2. *If e-implication is regular, then* $(\forall \beta \in dQ_1)(\forall \varphi \in \mathrm{Pres}Q)(\Gamma \cup \beta \vDash \varphi)$*.*

2.3.2 EVOCATION AND EROTETIC IMPLICATION

Both types of inferential structures can appear together. Let us investigate their interaction.

As it is demonstrated in the next example, e-implication does not preserve evocation. If we know $\Gamma \vDash Q_1$ and $Q_1 \vDash Q_2$, it does not mean that it must be $\Gamma \vDash Q_2$.

Example 6 (in CPL).

- $\{(\alpha \vee \beta)\} \vDash ?|\alpha, \beta|$ *and* $?|\alpha, \beta| \vDash ?(\alpha \vee \beta)$*, but* $\{(\alpha \vee \beta)\} \nvDash ?(\alpha \vee \beta)$*.*
- $\{\alpha\} \vDash ?|\alpha, \beta|$ *and* $\{\alpha\}, ?|\alpha, \beta| \vDash ?\alpha$*, but there is an answer to* $?\alpha$ *in* $\{\alpha\}$*.*

Of course, we do not see anything pathological in this example. It is superfluous to ask $?(\alpha \vee \beta)$, resp. $?\alpha$, when we know $(\alpha \vee \beta)$, resp. α.

Generally, this brings us back to the role of an auxiliary set of declaratives in erotetic implication. Due to the admissibility of *weakening in declaratives* (Fact 11), we can arrive at the structures of e-implications with Γ containing (direct) answers to some of the questions. On the other hand, there are some solutions of this problem proposed by erotetic logicians.[16]

In contrast to the previous example, we can prove that evocation is transferred through a regular e-implication.

Proposition 11. *If* $\Gamma \models Q_1$ *and* $Q_1 \vdash Q_2$, *then* $\Gamma \models Q_2$.

Proof. The first condition requires $\mathcal{M}^\Gamma \subseteq \bigcup_{\beta \in dQ_2} \mathcal{M}^\beta$. It is valid because of the same semantic range of both questions.

Let us suppose that there is $\beta \in dQ_2$ entailed by Γ. Then $\mathcal{M}^\Gamma \subseteq \mathcal{M}^\beta$ and, thanks to regularity of e-implication, $\mathcal{M}^\Gamma \subseteq \mathcal{M}^\alpha$, for some $\alpha \in dQ_1$. But it is in contradiction with $\Gamma \models Q_1$. □

Proposition 11 can be formulated not only in the version of pure regular e-implication:

Proposition 12. *If* $\Gamma \models Q_1$ *and* $\Gamma, Q_1 \vdash Q_2$, *then* $\Gamma \models Q_2$.

Proof. First, we prove $\Gamma \Vdash dQ_2$. Let us assume it is not true. Then there is the model $\mathbf{M}_0$ of Γ such that $\mathbf{M}_0 \not\models \beta$, for each $\beta \in dQ_2$. Because of $\Gamma \Vdash dQ_1$, there is $\alpha_0 \in dQ_1$ and $\mathbf{M}_0 \models \alpha_0$. From $\Gamma \cup \alpha \Vdash dQ_2$, for each $\alpha \in dQ_1$, there must be some $\beta_0 \in dQ_2$ such that $\mathbf{M}_0 \models \beta_0$ and that is a contradiction.

Second, let us assume that there is $\beta_0 \in dQ_2$ and $\Gamma \models \beta_0$. Regularity and the second condition of e-implication give $\Gamma \cup \beta_0 \models \alpha$ and it follows $\Gamma \models \alpha$ that is in contradiction with $\Gamma \models Q_1$. □

Since regularity was used only in the second part of the proof, we obtain an expected fact that $\Gamma \models Q_1$ and $\Gamma, Q_1 \vdash Q_2$ give soundness of an implied question Q_2 relative to Γ.[17]

At the first sight, it need not be $\Gamma, Q_1 \models Q_2$ or $\Gamma, Q_2 \models Q_1$ if we only know that $\Gamma \models Q_1$ as well as $\Gamma \models Q_2$.[18] Generally, neither evocation nor e-implication says something new about the structures of engaged questions. Nevertheless, we can expect that some clearing up of the structure of sets of direct answers

16 See, for example, the definition of *strong e-implication* given by Wiśniewski in [51]. Fact 13 includes the original inspiration for the definition of strong e-implication. The definition is the same as that of e-implication, but $\Gamma \not\Vdash \Delta$ is added in the second clause.

17 The second part of the proof of Proposition 12 could be slightly changed and we obtain that strong e-implication carries over as well. (Andrzej Wiśniewski called my attention to this.)

18 Let us take as an example (in CPL) the case that $\{\varphi\} \models ?\alpha$ and $\{\varphi\} \models ?\beta$. Then neither $\{\varphi\}, ?\alpha \models ?\beta$ nor $\{\varphi\}, ?\beta \models ?\alpha$.

could be helpful for the study of inferences. This will be discussed in the next section.

2.3.3 COMPARING QUESTIONS: RELATIONS BETWEEN QUESTIONS BASED ON DIRECT ANSWERS

It would be useful to be able to compare questions with respect to their 'answerhood power'. The chosen set-of-answers methodology brings us to a natural approach. We can either purely compare sets of direct answers or we can investigate their relationship based on entailment relation. Moreover, the cardinality of sets of direct answers can be controlled by a mapping from one set to the other.

Let us start with relations among questions based on a pure comparison of sets of direct answers.

Definition 10.

- *Two questions are* equal *($Q_1 = Q_2$) iff they have the same set of direct answers ($dQ_1 = dQ_2$).*[19]
- *A question Q_1 is* included *in a question Q_2 ($Q_1 \subset Q_2$) iff $dQ_1 \subset dQ_2$.*

This approach could be extended in a semantic way. We say that formula α *gives* an answer to a question Q iff there is $\beta \in dQ$ such that $\alpha \vDash \beta$. Having two questions Q_1 and Q_2 we can define a relationship of 'giving answers':

Definition 11. *A question Q_1* gives *a (direct) answer to Q_2 iff $(\forall \alpha \in dQ_1)(\exists \beta \in dQ_2)(\alpha \vDash \beta)$.*

In this definition the first question is considered to be (semantically) 'stronger' than the second one. For this we use the symbol $\geq$ and write $Q_1 \geq Q_2$.[20]

If $Q_1 = Q_2$ or $Q_1 \subset Q_2$, then $Q_1 \geq Q_2$ and, moreover, each direct answer to Q_1 not only gives an answer to Q_2 but also *is* a (direct) answer to Q_2, i.e., $(\forall \alpha \in dQ_1)(\exists \beta \in dQ_2)(\alpha \equiv \beta)$.

The ordering based on the relation $\geq$ has a slightly non-intuitive consequence: a completely contradictory question is the $\geq$-strongest one. However, the class of evoked questions is not affected by this problem.

Let us note an expected fact: $\geq$-stronger questions presuppose more than $\geq$-weaker ones.

Fact 17. *If $Q_1 \geq Q_2$, then* $\mathsf{Pres}Q_2 \subseteq \mathsf{Pres}Q_1$.

---

19 The original definition refers to *equivalent* questions instead of *equal* (cf. [51, p. 135]), but we use the first term for *erotetically equivalent* or semantically equivalent.

20 We will speak about ≥-stronger or ≥-weaker questions. The term *stronger* will be fixed in Definition 12.

This fact is not too useful. It is better to notice the relationship among maximal presuppositions. We have $\mathcal{M}^{\mathsf{Pres}Q_1} \subseteq \mathcal{M}^{\mathsf{Pres}Q_2}$. Each maximal presupposition of a $\geq$-stronger question entails a maximal presupposition of a $\geq$-weaker one; respectively, a prospective presupposition of a $\geq$-stronger question entails a prospective presupposition of a $\geq$-weaker question. The semantic range of a $\geq$-stronger question is included in the semantic range of a $\geq$-weaker question.

Fact 18. *If* $Q_1 \geq Q_2$, *then* $\bigcup_{\alpha \in dQ_1} \mathcal{M}^{\alpha} \subseteq \bigcup_{\beta \in dQ_2} \mathcal{M}^{\beta}$.

It follows that the safeness property is transmitted from $\geq$-stronger questions to $\geq$-weaker ones.

Fact 19. *If* Q_1 *is safe and* $Q_1 \geq Q_2$, *then* Q_2 *is safe.*

The next example shows that safeness of $\geq$-weaker questions is not transferred to $\geq$-stronger ones.

Example 7 (in CPL). $?\{\beta \wedge \alpha, \neg\beta\} \geq ?\beta$

Answerhood, evocation, and erotetic implication

The $\geq$-relation has an effect on inferences with questions. The first one is an obvious fact that an implied $\geq$-stronger question is regularly implied:

Proposition 13. *If* $\Gamma, Q_1 \models Q_2$ *and* $Q_2 \geq Q_1$, *then* $\Gamma, Q_1 \vdash Q_2$.

The proof is easy. Recall what is required in the definition of regular e-implication: $(\forall \beta \in dQ_2)(\exists \alpha \in dQ_1)(\Gamma \cup \beta \models \alpha)$.

If the relation $\geq$ is reverse, i.e., Q_1 gives an answer to Q_2, then whenever Q_1 implies Q_2, Q_2 regularly implies Q_1 (both with respect to Γ). Moreover, both questions are erotetically equivalent relative to Γ.

Proposition 14. *If* $\Gamma, Q_1 \models Q_2$ *and* $Q_1 \geq Q_2$, *then* $\Gamma, Q_2 \vdash Q_1$.

Proof. First, we need to show that $\Gamma \cup \beta \Vdash dQ_1$, for each $\beta \in dQ_2$. But it is an easy conclusion from $\Gamma, Q_1 \models Q_2$ because there is a subset $\Delta \subseteq dQ_1$ for each $\beta \in dQ_2$ such that $\Gamma \cup \beta \Vdash \Delta$.

The second condition of regular e-implication is obvious, it follows from $Q_1 \geq Q_2$. □

Generally, if Γ evokes Q_1 and Q_2, it need not be that either Q_1 implies Q_2 or Q_2 implies Q_1 (with respect to Γ). If a $\geq$-stronger question is evoked by Γ, then every $\geq$-weaker question regularly implies this stronger one with respect to Γ.

Proposition 15. *If* $\Gamma \models Q_1$ *and* $Q_1 \geq Q_2$, *then* $\Gamma, Q_2 \vdash Q_1$.

Proof. First, $\Gamma \cup \beta \Vdash dQ_1$ is required for each $\beta \in dQ_2$. We get $\Gamma \Vdash dQ_1$ from $\Gamma \models Q_1$. Now use the monotonicity of $\Vdash$.

Second, from $Q_1 \geq Q_2$ we have $(\forall\alpha \in dQ_1)(\exists\beta \in dQ_2)(\alpha \models \beta)$ and it gives the second condition of regular e-implication $(\forall\alpha \in dQ_1)(\exists\beta \in dQ_2)$ $(\Gamma \cup \alpha \models \beta)$. □

To digress for a moment, this repeated connection of $\geq$ and regular e-implication is not an accident. The definition of regular e-implication says that if $Q_1 \vdash Q_2$, then Q_2 gives an answer to Q_1, i.e., $Q_2 \geq Q_1$. However, 'giving an answer' does not produce e-implication, see the next example.

Example 8 (in CPL). *$?\{(\alpha\wedge\varphi),(\beta\wedge\psi)\} \geq ?\{\alpha,\beta\}$, but neither $?\{(\alpha\wedge\varphi),(\beta\wedge\psi)\} \models ?\{\alpha,\beta\}$ nor $?\{\alpha,\beta\} \models ?\{(\alpha \wedge \varphi),(\beta \wedge \psi)\}$.*

Let us return to evocation, it is clear that two equal questions are both evoked by a set of declaratives if one of them is evoked by this set. Generally, it is not sufficient to know $\Gamma \models Q_1$ and $Q_1 \geq Q_2$ to conclude $\Gamma \models Q_2$. An evoked $\geq$-stronger question only implies the soundness of $\geq$-weaker questions relative to Γ. Let us illustrate this in the case where the first question is included in the second one $(Q_1 \subset Q_2)$; there could be a direct answer to Q_2 which is entailed by Γ. This reminds us of the non-monotonic behavior of evocation. Notice that $Q_2 \subset Q_1$ will not help us either. In connection with the relation $\geq$ we have to require a version of equality. [21]

Fact 20. *Let $Q_1 \geq Q_2$ and $Q_2 \geq Q_1$. Then*

- *$\Gamma \models Q_1$ iff $\Gamma \models Q_2$,*
- *$Q_1 \vdash Q_2$ as well as $Q_2 \vdash Q_1$.*

Controlling the cardinality of sets of direct answers

The set of direct answers of a $\geq$-weaker question can be much larger than that of a $\geq$-stronger question. The book [51] introduces two relations originated from Tadeusz Kubiński that prevent this uncontrolled cardinality.

Definition 12. *A question Q_1 is* stronger *than Q_2 ($Q_1 \succeq Q_2$) iff there is a surjection $j : dQ_1 \rightarrow dQ_2$ such that for each $\alpha \in dQ_1$, $\alpha \models j(\alpha)$.*

The number of direct answers of the weaker question Q_2 does not exceed the cardinality of dQ_1, i.e., $|dQ_1| \geq |dQ_2|$. Above that, from the surjection, we know that each direct answer of a weaker question is given by some direct answer to a stronger question.

---

21 We do not introduce a special name for this relationship; it is included in the erotetic equivalence.

It is clear that if $Q_1 \succeq Q_2$, then $Q_1 \geq Q_2$. But, unfortunately, we cannot provide any special improvement of previous results for $\succeq$-relation. In particular, Examples 7 and 8 are valid for $\succeq$-relation as well.

Definition 13. *A question Q_1 is* equipollent *to a question Q_2 ($Q_1 \equiv Q_2$) iff there is a bijection $i : dQ_1 \rightarrow dQ_2$ such that for each $\alpha \in dQ_1$, $\alpha \equiv i(\alpha)$.*

In this case, both sets of direct answers have the same cardinality ($|dQ_1| = |dQ_2|$). Let us add expected results gained from equipollency.

Fact 21. *If $Q_1 \equiv Q_2$, then*

- *both $Q_1 \succeq Q_2$ and $Q_2 \succeq Q_1$,*
- *$\Gamma \models Q_1$ iff $\Gamma \models Q_2$,*
- *$Q_1 \vdash Q_2$ as well as $Q_2 \vdash Q_1$.*

Of course, two equal questions are equipollent.

Partial answerhood

We declared that the study of various types of answers (generally speaking, answerhood) is not the central point of this chapter. However, we can utilize the idea evoked by the second clause of the definition of erotetic implication (Definition 8). Narrowing down the set of direct answers of an implying question seems to be a good base for the definition of *partial answer.*

Definition 14. *A declarative φ* gives a partial answer *to question Q iff there is a non-empty proper subset $\Delta \subset dQ$ such that $\varphi \Vdash \Delta$.*[22]

This definition allows us to cover many terms from the concept of answerhood. Every direct answer gives a partial answer. Whenever ψ gives a (direct) answer, then ψ gives a partial answer. As a useful conclusion we obtain a weaker version of Proposition 15:

Fact 22. *If $\Gamma \models Q_1$ and each $\alpha \in dQ_1$ gives a partial answer to Q_2, then $\Gamma, Q_2 \models Q_1$.*

2.3.4 QUESTIONS AND SETS OF QUESTIONS

Working in classical logic, let us imagine we would like to know whether it is the case that α or it is the case that β. The question $?\{\alpha, \beta\}$ is posed. But there could be a problem when an entity, to which we are going to address this question, is not able to accept it. This can be caused, e.g., by a restricted language-acceptability, i.e., a device cannot 'understand' this question. However, assume

---

22 Compare this definition with Definition 4.10 in [51].

that there exist two devices such that: the first one can be asked by the question $?\alpha$, and the other one is able to work with the question $?\beta$. From both machines, independently, we get the following pairs of answers: $\{\alpha, \beta\}$, $\{\neg\alpha, \beta\}$, $\{\alpha, \neg\beta\}$ or $\{\neg\alpha, \neg\beta\}$.

Posing the question $Q = ?\{\alpha, \beta\}$ we know that if an answer to Q is true, then there must be a true answer to each question from the set $\{?\alpha, ?\beta\}$. Thus, we would like to require soundness transmission from an initial question to a set of questions.

Generally speaking, let us suppose that there are question Q and a set of questions $\Phi = \{Q_1, Q_2, ...\}$. For each model of a direct answer to Q there must be a direct answer in each Q_i valid in this model:

$$(\forall\alpha \in dQ)(\forall Q_i \in \Phi)(\alpha \Vdash dQ_i)$$

Possible states (of the world) given by answers to questions in the set Φ must be in a similar relation to the initial question. Whenever we keep a model of the choice of direct answers from each question in Φ, then there must be a direct answer to Q true in this model. For this, let us introduce a *choice function* ξ such that $\xi(Q_i)$ chooses exactly one direct answer from dQ_i. For each set of questions Φ and a choice function ξ there is a *choice set* $A_\xi^\Phi = \{\xi(Q_i) \mid Q_i \in \Phi\}$.[23] The soundness condition in the other direction (an initial question is sound with respect to each choice set) will be expressed, generally, by

$$(\forall A_\xi^\Phi)(A_\xi^\Phi \Vdash dQ).$$

Back to our example, there are four choice sets:

$$A_{\xi_1} = \{\alpha, \beta\}$$
$$A_{\xi_2} = \{\neg\alpha, \beta\}$$
$$A_{\xi_3} = \{\alpha, \neg\beta\}$$
$$A_{\xi_4} = \{\neg\alpha, \neg\beta\}$$

But the fourth one is not in compliance with the second soundness requirement, it is in contradiction with our (prospective) presupposition $(\alpha \vee \beta)$. If we admit the additional answer $(\neg\alpha \wedge \neg\beta)$ and a question in the form $?\{\alpha, \beta, (\neg\alpha \wedge \neg\beta)\}$, mutual soundness of this question and the set of questions $\{?\alpha, ?\beta\}$ will be valid. But this solution seems to be rather awkward. A questioner posing the question $?\{\alpha, \beta\}$ evidently presupposes $(\alpha \vee \beta)$. This will bring us to the definition of reducibility with respect to an auxiliary set of declaratives, and *mutual soundness* will be required in the following forms:

$$(\forall\alpha \in dQ)(\forall Q_i \in \Phi)(\Gamma \cup \alpha \Vdash dQ_i)$$

23 If the set Φ is clear from the context, we will write only A_ξ.

and

$$(\forall A^{\Phi}_{\xi})(\Gamma \cup A^{\Phi}_{\xi} \Vdash dQ).$$

Our example produces more than the soundness of Q relative to each A^{Φ}_{ξ} (with respect to Γ), also the *efficacy* of each A^{Φ}_{ξ} with respect to a question Q is valid:

$$(\forall A^{\Phi}_{\xi})(\exists \alpha \in dQ)(\Gamma \cup A^{\Phi}_{\xi} \vDash \alpha).$$

It will be reasonable to keep this strengthening. We are required to obtain at least one answer to an initial question from a choice set. Whenever Γ and A^{Φ}_{ξ} describe the state of the world, there must be a direct answer to a question Q that does the same job.

Reducibility of questions to sets of questions

We can take advantage of the previous discussion for the direct definition of reducibility of a question to a set of questions. Now, we introduce *pure reducibility* that does not use any auxiliary set of declaratives.

Definition 15. *A question Q is* purely reducible *to a non-empty set of questions Φ iff*

1. $(\forall \alpha \in dQ)(\forall Q_i \in \Phi)(\alpha \Vdash dQ_i)$
2. $(\forall A^{\Phi}_{\xi})(\exists \alpha \in dQ)(A^{\Phi}_{\xi} \vDash \alpha)$
3. $(\forall Q_i \in \Phi)(|dQ_i| \leq |dQ|)$

The first two conditions express mutual soundness, the second one adds efficacy, as it was discussed, and the last one requires *relative simplicity*. If Q is reducible to a set Φ, we will write $Q \gg \Phi$. The definition of pure reducibility was introduced by Andrzej Wiśniewski in [50].

Example 9 (in CPL).

- $?\{\alpha, \beta, (\neg\alpha \wedge \neg\beta)\} \gg \{?\alpha, ?\beta\}$
- $?|\alpha, \beta| \gg \{?\alpha, ?\beta\}$
- $?(\alpha \circ \beta) \gg \{?\alpha, ?\beta\}$, *where $\circ$ is any of the connectives: $\wedge, \vee, \rightarrow$*

In the first item, there is the 'pure' version from the introductory discussion. All items display reducibilities between initial safe questions and sets of safe questions. The following proposition shows that it is not an accident.

Proposition 16. *If $Q \gg \Phi$, then Q is safe iff each $Q_i \in \Phi$ is safe,*

Proof. The first condition of Definition 15 can be rewritten as $\bigcup_{\alpha \in dQ} \mathcal{M}^{\alpha} \subseteq \bigcup_{\beta \in dQ_i} \mathcal{M}^{\beta}$, for each $Q_i \in \Phi$, and it gives the implication from left to right.

For proof of the other implication, let us suppose $Q \gg \Phi$ and that each $Q_i \in \Phi$ is safe, but Q is not. It implies the existence of model $\mathbf{M}_0 \in \mathcal{M}$ such that $\mathbf{M}_0 \nvDash$

α, for each $\alpha \in dQ$. The safeness of all Q_i gives $(\forall Q_i \in \Phi)(\exists \beta \in dQ_i)(\mathbf{M}_0 \models \beta)$. Thus, there is A_ξ^Φ made from these βs and $\mathbf{M}_0 \models A_\xi^\Phi$.[24] But it is in contradiction with the second condition of the definition of $Q \gg \Phi$, which gives the existence of some $\alpha \in dQ$ such that $\mathbf{M}_0 \models \alpha$. □

From this we know that if there is a risky question among questions in Φ and $Q \gg \Phi$, then Q must be risky too (cf. [51, p. 197]).

The rewritten first condition of Definition 15 is of the form

$$\bigcup_{\alpha \in dQ} \mathcal{M}^\alpha \subseteq \bigcap_i \bigcup_{\beta \in dQ_i} \mathcal{M}^\beta$$

and it brings out the relationship of semantic ranges. The semantic range of a reduced question is bounded by the intersection of all semantic ranges of Q_is.

The relation of pure reducibility is 'reflexive' ($Q \gg \{Q\}$) and we can prove the following version of transitivity:[25]

Proposition 17. *If $Q \gg \Phi$ and each $Q_i \in \Phi$ is reducible to some set of questions Φ_i, then $Q \gg \bigcup_i \Phi_i$.*

Proof. The third condition of Definition 15 is clearly valid.

The first one is easy to prove. From $Q \gg \Phi$ we get $\bigcup_{\alpha \in dQ} \mathcal{M}^\alpha \subseteq \bigcup_{\beta \in dQ_i} \mathcal{M}^\beta$, for each $Q_i \in \Phi$, and from the reducibility of all Q_i in Φ to an appropriate Φ_i we have $\bigcup_{\beta \in dQ_i} \mathcal{M}^\beta \subseteq \bigcup_{\gamma \in dQ_j} \mathcal{M}^\gamma$, for each $dQ_j \in \Phi_i$. It gives $\bigcup_{\alpha \in dQ} \mathcal{M}^\alpha \subseteq \bigcup_{\gamma \in dQ_j} \mathcal{M}^\gamma$, for each $Q_j \in \bigcup_i \Phi_i$.

For the second one, we require for each $A_\xi^{\bigcup_i \Phi_i}$ the existence of $\alpha \in dQ$ such that $A_\xi^{\bigcup_i \Phi_i} \models \alpha$. From the reducibility of all Q_i in Φ to an appropriate Φ_i we have that each $A_{\xi'}^{\Phi_i}$ is a subset of some $A_\xi^{\bigcup_i \Phi_i}$. It implies that if there is any model $\mathbf{M}$ of $A_\xi^{\bigcup_i \Phi_i}$, it must be a model of some $A_{\xi'}^{\Phi_i}$. From $Q \gg \Phi$ we know that there is $\alpha \in dQ$ for each choice set $A_{\xi''}^\Phi$ on Φ. This choice set is made by elements of all $dQ_i \in \Phi$ which are valid in $\mathbf{M}$. It means that $A_{\xi''}^\Phi$ is valid in $\mathbf{M}$ as well as $\alpha \in dQ$. □

Now let us look at the relationship of pure reducibility and pure e-implication. The following example shows that it need not be that e-implication causes reducubility. Both definitions have the same first conditions, but the second condition of reducibility can fail.

Example 10 (in CPL). $?|\alpha, \beta| \models ?(\alpha \wedge \beta)$, *but* $?|\alpha, \beta| \not\gg \{?(\alpha \wedge \beta)\}$.

---

24 We have to be careful to say this in the general case where Φ or each dQ_i are infinite; axiom of choice can be required.

25 Proposition 17 corresponds to Corollary 7.6 in [51].

On the other hand, we can prove that regular e-implication implies reducibility.

Proposition 18. *Let Φ be a set of questions such that $Q \vdash Q_i$, for each $Q_i \in \Phi$. If $(\forall Q_i \in \Phi)(|dQ_i| \leq |dQ|)$, then $Q \gg \Phi$.*

Proof. Let us prove the second condition of Definition 15 that requires for each A^{Φ}_{ξ} the existence of $\alpha \in dQ$ such that $\mathcal{M}^{A^{\Phi}_{\xi}} \subseteq \mathcal{M}^{\alpha}$. It is known that $\mathcal{M}^{A^{\Phi}_{\xi}} \subseteq \mathcal{M}^{\beta}$, for each $\beta \in A^{\Phi}_{\xi}$. If $Q \vdash Q_i$, then for each $\beta \in dQ_i$ there is $\alpha \in dQ$ such that $\mathcal{M}^{\beta} \subseteq \mathcal{M}^{\alpha}$. Thus, $\mathcal{M}^{A^{\Phi}_{\xi}} \subseteq \mathcal{M}^{\alpha}$. □

What about if we know $Q_i \models Q$ or, even, $Q_i \vdash Q$, for each $Q_i \in \Phi$, and $(\forall Q_i \in \Phi)(|dQ_i| \leq |dQ|)$, can we conclude that $Q \gg \Phi$? Example 10 gives the negative answer to this question. Moreover, let us notice that $?(\alpha \wedge \beta) \vdash ?|\alpha, \beta|$.

Not even reducibility produces e-implication.

Example 11 (in CPL). *$?(\alpha \wedge \beta) \gg \{?\alpha, ?\beta\}$, but $?(\alpha \wedge \beta)$ is not implied either by $?\alpha$ or by $?\beta$, and $?(\alpha \wedge \beta)$ does not imply neither $?\alpha$ nor $?\beta$.*

In the next subsection we will study some special cases of links between reducibility and e-implication.

So far we have worked only with pure reducibility. It could be useful to introduce the general term of *reducibility* with respect to a context given by a set of declaratives. The definition is almost the same as Definition 15, but the mutual soundness and efficacy conditions are supplemented by an auxiliary set of declaratives Γ (cf. [23]). We will write $\Gamma, Q \gg \Phi$.

Definition 16 (Reducibility). *A question Q is* reducible *to a non-empty set of questions Φ with respect to a set of declaratives Γ iff*

1. $(\forall \alpha \in dQ)(\forall Q_i \in \Phi)(\Gamma \cup \alpha \Vdash dQ_i)$
2. $(\forall A^{\Phi}_{\xi})(\exists \alpha \in dQ)(\Gamma \cup A^{\Phi}_{\xi} \models \alpha)$
3. $(\forall Q_i \in \Phi)(|dQ_i| \leq |dQ|)$

The introductory example of this section with respect to the discussion we have made is recalled in this example:

Example 12 (in CPL). $(\alpha \vee \beta), ?\{\alpha, \beta\} \gg \{?\alpha, ?\beta\}$

As it is expected, the role of Γ is similar to the role of an auxiliary set of declaratives in e-implication. Γ provides an appropriate context.

Fact 23. *If $Q \gg \Phi$, then $\Gamma, Q \gg \Phi$, for each Γ.*

So, we can speak of *weakening in declaratives* and it enables us to generalize Proposition 18.

Proposition 19. *If* $\Gamma, Q \vdash Q_i$*, for each* $Q_i \in \Phi$*, and* $(\forall Q_i \in \Phi)(|dQ_i| \leq |dQ|)$*, then* $\Gamma, Q \gg \Phi$*.*

Proof. The third and the first conditions of Definition 16 are obvious.

The second one requires that for each A_ξ^Φ there is $\alpha \in dQ$ such that $\mathcal{M}^{\Gamma \cup A_\xi^\Phi} \subseteq \mathcal{M}^\alpha$. From the construction of choice sets we know that for each A_ξ^Φ and $Q_i \in \Phi$ there is $\beta \in dQ_i$ (member of A_ξ^Φ) such that $\mathcal{M}^{\Gamma \cup A_\xi^\Phi} \subseteq \mathcal{M}^{\Gamma \cup \beta}$. The regular e-implication provides that there is $\alpha \in dQ$ for each $\beta \in dQ_i$ such that $\mathcal{M}^{\Gamma \cup \beta} \subseteq \mathcal{M}^\alpha$. □

We close this subsection by reversing the 'direction' of the reducibility relation. Let us suppose that we have generated a set of questions Φ that are evoked by a set of declaratives Γ. Can we conclude that Γ evokes such a complex question which is reducible to the set Φ? Generally, not. But if we know that the complex question gives an answer to some question from Φ, the answer will be positive.

Proposition 20. *If* Γ *evokes each question from a set* Φ*,* $Q \gg \Phi$*, and there is a question* $Q_i \in \Phi$ *such that* $Q \geq Q_i$*, then* $\Gamma \models Q$*.*

Proof. Soundness of Q relative to Γ requires for each model $\mathbf{M} \models \Gamma$ the existence of an answer $\alpha \in dQ$ true in it. From the evocation of each $Q_i \in \Phi$ we have $(\forall \mathbf{M} \models \Gamma)(\forall Q_i \in \Phi)(\exists \beta \in dQ_i)(\mathbf{M} \models \beta)$. So, each model of Γ produces some choice set such that $(\forall \mathbf{M} \models \Gamma)(\exists A_\xi^\Phi)(\mathbf{M} \models A_\xi^\Phi)$. Together with reducibility, where it is stated that $(\forall A_\xi^\Phi)(\exists \alpha \in dQ)(A_\xi^\Phi \models \alpha)$, we get $\Gamma \Vdash dQ$.

Informativness of Q with respect to Γ is justified by $\geq$-relation for some question $Q_i \in \Phi$. If $\Gamma \models \alpha$, for some $\alpha \in dQ$, then it creates a contradiction with $\Gamma \models Q_i$. □

Reducibility and sets of yes-no questions

The concept of reducibility is primarily devoted to a transformation of a question to a set of 'less complex' questions. The introductory discussion and its formalization in Example 12 evoke interesting questions:

- If we have an initial question $Q = ?\{\alpha_1, \alpha_2, ...\}$ with, at worst, a countable list of direct answers, is it possible to reduce it to a set of yes-no questions based only on the direct answers of Q?
- Moreover, could we require the e-implication relationship between Q and questions in the set Φ?

We can find an easy solution to these problems under the condition that yes-no questions are safe and we have an appropriate set of declaratives—we will require Q to be sound with respect to Γ.

Proposition 21. *Let us suppose that yes-no questions are safe in the background logic. If question $Q = ?\{\alpha_1, \alpha_2, ...\}$ is sound with respect to a set Γ, then there is a set of yes-no questions Φ such that $\Gamma, Q \gg \Phi$ and $\Gamma, Q \models Q_i$, for each $Q_i \in \Phi$.*

Proof. First, we define the set of yes-no questions Φ based on the initial question $Q = ?\{\alpha_1, \alpha_2, ...\}$ such that

$$\Phi = \{?\alpha_1, ?\alpha_2, ...\}.$$

Second, we prove the condition that is common for both reducibility and e-implication. The safeness of members of Φ implies that $\mathcal{M}^{\alpha} \subseteq \bigcup_{\beta \in dQ_i} \mathcal{M}^{\beta}$, for each $\alpha \in dQ$ and $Q_i \in \Phi$. This gives $\mathcal{M}^{\Gamma \cup \alpha} \subseteq \bigcup_{\beta \in dQ_i} \mathcal{M}^{\beta}$.

To prove reducibility we have to justify the second condition of Definition 16. We need to find an $\alpha \in dQ$ for each A^{Φ}_{ξ} such that $\Gamma \cup A^{\Phi}_{\xi} \models \alpha$. Two cases will be distinguished.

1. If there is α from both A^{Φ}_{ξ} and dQ, then choose this direct answer.
2. If there is no direct answer $\alpha \in dQ$ in A^{Φ}_{ξ}, then $\mathcal{M}^{\Gamma \cup A^{\Phi}_{\xi}} = \emptyset$ and we can take any α from dQ.

The final step is the proof of e-implication. We have to show that for each Q_i and each direct answer $\beta \in dQ_i$ there is a non-empty subset $\Delta \subset dQ$ such that $\Gamma \cup \beta \Vdash \Delta$. For this, we use the form of questions in Φ.

1. If $\beta \in dQ$, then Δ could be $\{\beta\}$ and $\Gamma \cup \beta \Vdash \{\beta\}$.
2. If $\beta \notin dQ$ and $\Gamma \cup \beta$ has at least one model, then $\mathcal{M}^{\Gamma \cup \beta} \subseteq \mathcal{M}^{\Gamma}$. Simultaneously, β must be of the form $\neg\alpha_j$ and Δ can be defined as $dQ \setminus \{\alpha_j\}$. Together with soundness of the initial question Q with respect to Γ, which means $\mathcal{M}^{\Gamma} \subseteq \bigcup_{\alpha \in dQ} \mathcal{M}^{\alpha}$, we get $\mathcal{M}^{\Gamma \cup \beta} \subseteq \bigcup_{\alpha \in \Delta} \mathcal{M}^{\alpha}$.

□

This proposition enables us to work with classes of questions which are known to be sound relative to the sets of their presuppositions. Normal and regular questions are the obvious examples. Whenever we know that the initial question is evoked by a set of declaratives, we obtain the following conclusion.

Fact 24. *Working in logic where yes-no questions are safe, if a question $Q = ?\{\alpha_1, \alpha_2, ...\}$ is evoked by a set of declaratives Γ, then there is a set of yes-no questions Φ such that $\Gamma, Q \gg \Phi$ and $\Gamma, Q \models Q_i$, for each $Q_i \in \Phi$.*

This fact corresponds to the main lemma in the paper [23, pp. 104–5] where a bit of a different definition of reducibility is used, but the results are the same.

If dQ is finite or the entailment is compact, the set Φ is a finite set of yes-no questions. Simultaneously, it is useful to emphasize that the proof of Proposition 21 shows how to construct such a set.[26]

---

26 The same result is provided by theorems 7.49–7.51 in [51].

In logic with risky yes-no questions, the first condition of reducibility as well as e-implication can fail. It need not be $\mathcal{M}^{\Gamma,\alpha} \subseteq (\mathcal{M}^{\beta} \cup \mathcal{M}^{\neg\beta})$, for each $\alpha \in dQ$ and each $?\beta \in \Phi$. We can ask for the help of an auxiliary set of declaratives again. Let us remember Fact 13 and add the soundness of each $Q_i \in \Phi$ with respect to Γ. Reviewing the proof of Proposition 21, the rest is valid independently of the safeness of yes-no questions. As a corollary we have:

Fact 25. *If a question $Q = ?\{\alpha_1, \alpha_2, ...\}$ is sound relative to a set Γ, and there is a set of yes-no questions $\Phi = \{?\alpha_1, ?\alpha_2, ...\}$ (based on Q) such that $\Gamma \Vdash dQ_i$, for each $Q_i \in \Phi$, then $\Gamma, Q \gg \Phi$ and $\Gamma, Q \vDash Q_i$, for each $Q_i \in \Phi$.*

The construction of yes-no questions provided by Proposition 21 does not prevent a high complexity of such yes-no questions. Observing the last item of Example 9, it seems worthwhile to inquire whether it is possible to follow this process and to reduce a question (with respect to an auxiliary set of declaratives) to a set of atomic yes-no questions based on subformulas of the initial question (cf. [57]). The first restriction is clear, yes-no questions must be safe. But that is not all, the second clause of pure reducibility (Definition 15) requires the *truth-functional* connection of subformulas. Then the answer is positive. We can repeatedly use a cautious extension of Proposition 17:

Fact 26. *If $\Gamma, Q \gg \Phi$ and each $Q_i \in \Phi$ is reducible to some set of questions Φ_i, then $\Gamma, Q \gg \bigcup_i \Phi_i$.*

There is a similar concept in literature, called *erotetic search scenarios*, based on properties of the classical logic and erotetic implication (see [53, 54, 55] and [45]). The idea is that there is an initial question (and a context) and we can reach an answer to it by searching through the answers of some operative (auxiliary) questions. Scenarios are trees, an initial question (and a context) is the root and branching is based on direct answers of auxiliary questions. The relationship between interrogative nodes is given by erotetic implication. Some scenarios work with the descending 'complexity' of questions from the root to leafs. If we recall the non-transitivity of e-implication (Example 5), we can recognize the 'truth-functional' auxiliary role of the question $?|\alpha, \beta|$ as an interlink between $?(\alpha \wedge \beta)$ and questions $?\alpha$ and $?\beta$.

2.4 INFERENTIAL EROTETIC LOGIC (FINAL REMARKS)

Inferential erotetic logic has been introduced as a very general theory of erotetic inferences dealing with a general language $\mathcal{L}$ extended by questions. Following the IEL philosophy and in accordance with our first chapter, we decided to introduce questions as sets of declaratives, which are in the role of direct answers. Inferences with questions are based on multiple-conclusion entailment that makes it possible to work with sets of declaratives in premises as well as in conclusions. Thus, questions are kept as special objects of the language $\mathcal{L}_Q$, they are not combined by logical constants as declaratives are. All relationships among questions are based on inferences and a comparison of their sets of direct answers. The main modification of the original IEL can bee seen in the chosen SAM and in the model-based approach; the term *semantic range* of a question made the semantic work easier and more transparent.

In this chapter we studied many properties and relationships, but the central point was whether we could formulate some relationships (meta-rules) for erotetic consequence relations in IEL. Our general view did not tend to provide any axiomatization. The properties and relationships discussed vary with the chosen background system, especially, they depend on semantics. IEL opens up many possibilities for working with questions in various logical systems (see [7] as a nice example) and serves as an inspiration for work in erotetic logic.

3. EPISTEMIC LOGIC WITH QUESTIONS

3.1 INTRODUCTION

Communication is essentially connected with the exchange of information. One of the basic ways to complete someone's knowledge is the posing of questions. Although questions differ from declaratives, they play a similar role in reasoning. In the recent history of erotetic logic we have seen success in finding a desirable position of formal approaches to questions and in the study of inferences based on them. Let us recall Wiśniewski's inferential erotetic logic and the intensional approach of Groenendijk and Stokhof. Such approaches made it possible to incorporate questions into some formal systems and not to lose their specific position in inferences. This brings us to an important point. On the one hand, questions are specific entities and they bear some special properties. On the other hand, they are used in formal systems that focus on reasoning.

In Section 1.2 we introduced a formalization of questions called set-of-answers methodology. It utilizes the close connection of questions and their answerhood conditions. Simultaneously, we mentioned there that it is very natural to use epistemic terms in speaking of questions.

The approach we are going to present in this chapter works with epistemic logic as a background. Roughly speaking, epistemic logic and its semantics are used for the modeling of knowledge and epistemic possibilities of an agent, as well as groups of agents. Epistemic logic in use will be the normal multi-modal logic. At the very beginning we do not impose any conditions on it; thus, the working system will be the normal modal (epistemic) logic K with its standard relational semantics (Kripke frames and models). The way of incorporating questions should be of such generality that it could be applied in many epistemic-like logic systems.

The process of communication is a dynamic matter and epistemic logic is open to enrichment by dynamic aspects, see, e.g., [48]. These aspects will be studied later on in Chapter 4 where we accept the 'standard' model of knowledge based on modal system S5. Now let us briefly sum up what a question and its epistemic meaning are.

In the introductory chapter we used the example with three friends holding cards and Catherine wanting to find the Joker-card holder. She asks

Who has the Joker: Anne, or Bill?

From the viewpoint of SAM, the question has a two-element set of direct answers:

{Anne has the Joker, Bill has the Joker}

Asking this question, Catherine expresses that she

1. does not know the right answer to the question,
2. considers the answers to be possible, and, moreover,
3. presupposes that either Anne has the Joker or Bill has it.

An agent-questioner provides the information of her ignorance (item 1) as well as the expected way to complete her knowledge: 2 says that there are two possibilities, and 3 that one of the possibilities is expected. A question provides information about the epistemic state of the agent who asks the question. An asked question means that listeners can form a partial picture of the questioner's knowledge structure, which is an important part in communication and the solving of problems in groups.[1] As it was mentioned at the beginning, an exchange of information is the basis of communication. The main aim of communication in a group is to share data and to solve problems. A very typical example is a group of scientists trying to find an answer to their scientific problem. They can reach a solution only by sharing their knowledge and ignorance.

Communication will be studied in the next chapter, here we prepare an 'erotetic epistemic framework'. First, we introduce propositional single-agent (normal modal) epistemic logic and extend it trough questions. We fully apply our set-of-answers methodology and allow the mix of declaratives and interrogatives in sets of direct answers. Questions will be a natural part of inference relations based on the background logic. Then we discuss answerhood conditions in a relationship with conditions imposed on a question. And, finally, the role of epistemic context as well as sets of questions are studied.

---

1 Jeroen Groenendijk says that "assertions may provide new *data*, questions may provide new *issues*" [14].

3.2 SINGLE-AGENT PROPOSITIONAL EPISTEMIC LOGIC AND QUESTIONS

Our approach starts with the language of single-agent propositional epistemic logic. The language of classical propositional logic $\mathcal{L}_{cpl}$ is extended by the modalities $[i]$ and $\langle i \rangle$. The former modality can be interpreted as 'agent i knows that ...', 'agent i believes that ...', etc. The latter is an 'epistemic possibility', it can be interpreted as 'agent i considers ... possible'. We obtain a language $\mathcal{L}^{K}_{cpl}$ with a subset of signs for atomic formulas $\mathcal{P} = \{p, q, ...\}$ and formulas defined as follows:

$$\varphi ::= p \mid \neg\psi \mid \psi_1 \vee \psi_2 \mid \psi_1 \wedge \psi_2 \mid \psi_1 \rightarrow \psi_2 \mid \psi_1 \leftrightarrow \psi_2 \mid [i]\psi \mid \langle i \rangle\psi$$

In multi-agent variants of epistemic logic we presuppose that there is a finite set of agents $\mathcal{A} = \{1, \dots, m\}$, where numbers $1, \dots, m$ are names for agents. This section deals with a single-agent variant for the sake of simplicity in introducing questions and their basic properties. Moreover, we do not work with the interpretation of $[i]$ in terms 'knowledge' or 'belief' (of an agent i), it is just the 'epistemic necessity' of an agent without restrictions to knowledge or belief conditions.[2]

Semantics is based on Kripke-style models for normal modal logics. A *Kripke frame* is a relational structure $\mathcal{F} = \langle S, R_i \rangle$ with a set of states (points, indices, possible worlds), S and an accessibility relation $R_i \subseteq S^2$ for (every) $i \in \mathcal{A}$. The *Kripke model* $\mathbf{M}$ is a pair $\langle \mathcal{F}, v \rangle$ where v is a valuation of atomic formulas. The satisfaction relation $\models$ is defined in the standard way:

1. $(\mathbf{M}, s) \models p$ iff $s \in v(p)$
2. $(\mathbf{M}, s) \models \neg\varphi$ iff $(\mathbf{M}, s) \not\models \varphi$
3. $(\mathbf{M}, s) \models \psi_1 \vee \psi_2$ iff $(\mathbf{M}, s) \models \psi_1$ or $(\mathbf{M}, s) \models \psi_2$
4. $(\mathbf{M}, s) \models \psi_1 \wedge \psi_2$ iff $(\mathbf{M}, s) \models \psi_1$ and $(\mathbf{M}, s) \models \psi_2$
5. $(\mathbf{M}, s) \models \psi_1 \rightarrow \psi_2$ iff $(\mathbf{M}, s) \models \psi_1$ implies $(\mathbf{M}, s) \models \psi_2$
6. $(\mathbf{M}, s) \models [i]\varphi$ iff $(\mathbf{M}, s_1) \models \varphi$, for each s_1 such that sR_is_1

Modality $\langle i \rangle$ is understood to be dual to $[i]$:

$$\langle i \rangle\varphi \equiv \neg[i]\neg\varphi$$

We do not put any restrictions on accessibility relation now, thus, we have semantics for the modal system K.

---

2 That is the reason for the use of $[i]$ instead of the signs K_i or B_i. We use K only as a superscript in the language $\mathcal{L}^{K}_{cpl}$.

3.2.1 INCORPORATING QUESTIONS

We extend the epistemic language $\mathcal{L}^K_{cpl}$ with brackets $\{,\}$ and the question mark $?_i$ for a question asked by an agent i. The obtained language is $\mathcal{L}^{KQ}_{cpl}$. Metavariables Q^i, Q^i_1, etc., will be used for interrogative formulas.

A question Q^i is any formula of the form

$$?_i\{\alpha_1, \dots, \alpha_n\},$$

where $dQ^i = \{\alpha_1, \dots, \alpha_n\}$ is the set of direct answers to a question Q. Direct answers are formulas of our extended epistemic language $\mathcal{L}^{KQ}_{cpl}$ and questions can be among direct answers as well.[3] We suppose that dQ^i is finite with at least two syntactically distinct elements. In accordance with our set-of-answers methodology, the intended reading of a question Q^i is:

> An agent i is asking: Is it the case that α_1 or is it the case that α_2 ... or is it the case that α_n?

Whenever she asks such question, she presupposes that at least one of the direct answers is the case. Whenever someone hears such a question, she knows that a questioner presupposes the same, i.e., at least one of the direct answers is the case. This brings us to an important term *presupposition*, which is studied in the next subsection. On the contrary to our liberal SAM, we require dQ^i to be finite. Working in propositional logic, we want to keep direct answers as clear epistemic possibilities, this seems to be useful in communication processes.

Presuppositions

Taking inspiration from *inferential erotetic logic* (see Section 2.2.2), we define presuppositions of questions as formulas that are implied by each direct answer. A presupposition is a 'consequence' of each direct answer, no matter which answer is right.[4]

Definition 17. *A formula φ is a* presupposition *of a question Q iff $(\alpha \to \varphi)$ is valid for each $\alpha \in dQ$. We write $\varphi \in \mathsf{Pres}Q$.*

The set of presuppositions of a question is full of redundant formulas. To avoid this let us define a class of *maximal presuppositions*. Maximal presuppositions imply every presupposition.

Definition 18. *A formula φ is a* maximal presupposition *iff $\varphi \in \mathsf{Pres}Q$ and $(\varphi \to \psi)$ is valid for each $\psi \in \mathsf{Pres}Q$.*

---

3 Compare it with IEL's requirement that $\alpha_1, \dots, \alpha_n$ are only declarative formulas, see Section 2.1.1.

4 Let us make a symbol convention: if it is not necessary to use the index i, we will omit it.

Example 13. *A formula* $(\alpha_1 \vee \ldots \vee \alpha_n)$ *is a maximal presupposition of a question* $?\{\alpha_1, \ldots, \alpha_n\}$.

The theory of questions in IEL introduces one more term—*prospective presupposition*. The following definition is inspired by a modal reformulation of the original IEL definition of prospective presuppositions: The truth of a prospective presupposition in the state of a model gives the truth of some direct answer in this state.

Definition 19. *A formula* φ *is a* prospective presupposition *of a question* Q *iff* $\varphi \in \mathsf{Pres}Q$ *and, for all models* $\mathbf{M}$ *and states* s, *if* $(\mathbf{M}, s) \models \varphi$, *then there is a direct answer* $\alpha \in dQ$ *such that* $(\mathbf{M}, s) \models \alpha$. *We write* $\varphi \in \mathsf{PPres}Q$.

A formula $(\alpha_1 \vee \ldots \vee \alpha_n)$ is a prospective presupposition of a question $?\{\alpha_1, \ldots, \alpha_n\}$ as well.

Things are much easier when working with finite sets of direct answers and being in a system that extends classical propositional logic. We need not distinguish between maximal and prospective presuppositions.[5]

Proposition 22. *The set of prospective presuppositions is equal to the set of maximal presuppositions of a question* Q.

Proof. First, let $\varphi \in \mathsf{PPres}Q$ but assume that φ is not maximal. Since φ is not maximal, there must be $(\mathbf{M}, s)$ and $\psi \in \mathsf{Pres}Q$ such that $(\mathbf{M}, s) \models \varphi$ and $(\mathbf{M}, s) \not\models \psi$. $(\mathbf{M}, s) \models \varphi$ implies the existence of a direct answer α satisfied in $(\mathbf{M}, s)$. Each presupposition is implied by every direct answer, so is ψ in $(\mathbf{M}, s)$ and it gives $(\mathbf{M}, s) \models \psi$, which is a contradiction.

Second, let φ be maximal, but not prospective. In our (finite) case we can suppose that Q has at least one prospective presupposition. If φ is not prospective, then there is the state $(\mathbf{M}, s) \models \varphi$ and $(\mathbf{M}, s) \not\models \alpha$, for each $\alpha \in dQ$. All presuppositions are satisfied in the state $(\mathbf{M}, s)$, so are prospective presuppositions, but it is in contradiction with the fact that no α holds in $(\mathbf{M}, s)$. □

In the next proposition we show the same result we have obtained for IEL (see Proposition 2), namely, that all prospective presuppositions of a question are semantically equivalent.

Proposition 23. *If* $\varphi, \psi \in \mathsf{PPres}Q$, *then* $\varphi \equiv \psi$.

Proof. For proving semantic equivalence we have to prove $\varphi \models \psi$ as well as $\psi \models \varphi$.

If $(\mathbf{M}, s) \models \varphi$, then there is $\alpha \in dQ$ such that $(\mathbf{M}, s) \models \alpha$. Since $\psi \in \mathsf{Pres}Q$, then $(\mathbf{M}, s) \models (\alpha \to \psi)$ gives $(\mathbf{M}, s) \models \psi$. We have obtained $\varphi \models \psi$.

The other case is similar. □

---

5 In IEL, prospective presuppositions are maximal, but not vice versa, see [51, Corollary 4.10].

Thus, the symbol PPres$Q$ will be used for a formula representing the prospective presuppositions of a question Q modulo the semantic equivalence.

Note on presupposing and context In the example with card players we mentioned Catherine's question

Who has the Joker: Anne, or Bill?

In item 3 (see p. 54) we wrote that *either Anne or Bill has the Joker* is Catherine's presupposition, i.e., it must be the case that Anne has the Joker (and Bill not), or it must be the case that Bill has the Joker (and Anne not). This is indicated by the comma in the interrogative sentence as well. Catherine's presupposition is under the influence of the context given by the rules of the card 'game':

> Just one Joker is distributed among the agents Anne, Bill, and Catherine.

Now, let us pose the following question:

What is Peter: a lawyer or an economist?

If there is no supplementary context, the question bears a presupposition that Peter is at least one of the two possibilities (maybe, both of them). The role of context will be studied later on, viz. subsection Relativized askability in 3.2.2, especially.

Askable questions

In semantics for a majority of logical systems we speak about the truth or falsity of a formula (in a particular state of a particular model). It is clear that it makes little sense to speak about the truth/falsity of a question. We introduce instead a concept of *askability* of a question. Askability is based on our idea of a 'reasonable' question in a certain situation. 'Reasonability' corresponds to the three conditions we informally mentioned at the introductory example, see page 54. Let us repeat them and give them names:

1. **Non-triviality** It is not reasonable to ask a question if the answer is known.
2. **Admissibility** Each direct answer is considered as possible.
3. **Context** At least one of the direct answers must be the right one.

Whenever an agent-questioner poses a question, she does not know which (direct) answer to a question is true, but, simultaneously, she considers all (direct) answers possible and she is aware of what is presupposed—she knows the prospective presupposition of a question. The formal definition follows.

Definition 20 (Askability). *It holds for a question* $Q^i = ?_i\{\alpha_1, \dots, \alpha_n\}$ *that*

$$(\mathbf{M}, s) \vDash Q^i$$

iff

1. $(\mathbf{M}, s) \nvDash [i]\alpha$, *for each* $\alpha \in dQ^i$
2. $(\mathbf{M}, s) \vDash \langle i \rangle\alpha$, *for each* $\alpha \in dQ^i$
3. $(\mathbf{M}, s) \vDash [i]\mathsf{PPres}Q^i$

Then we say that Q^i *is* askable *in the state* $(\mathbf{M}, s)$ *(by an agent i).*

As we can see, the freedom of the syntactical form of questions was partly compensated by restrictions in semantics. We say that a question is (generally) *askable* iff there is a model and a state where the question is askable (by an agent). Askable questions include neither contradiction nor tautology among their direct answers. The former is excluded by the second condition and the latter by the first one. A question Q^i is *askable (by an agent i) relative to a model* $\mathbf{M}$ (let us write $\mathbf{M} \vDash Q^i$) iff $(\mathbf{M}, s) \vDash Q^i$ for each $s \in S$ of the model $\mathbf{M}$. The definition of $\vDash Q^i$ is straightforward, but there are no 'tautological' questions in K. A question is not askable in a state without successors. In our version, at least two successors are needed. If we work in systems extending classical logic, the first condition is equal to $(\mathbf{M}, s) \vDash \neg[i]\alpha$, i.e., $(\mathbf{M}, s) \vDash \langle i \rangle\neg\alpha$, for each $\alpha \in dQ^i$. We can understand the questioner as admitting the possibility of $\neg\alpha$ for each direct answer α to a question Q^i. In these systems, questions are complex modal formulas. However, Definition 20 is meant in full generality without the intention to reduce questions to any epistemic formulas.

The epistemic semantic viewpoint represents the agent's 'knowledge' in a state s as an afterset sR_i given by the states related to s by an accessibility relation R_i, i.e., $sR_i = \{s' : sR_is'\}$.

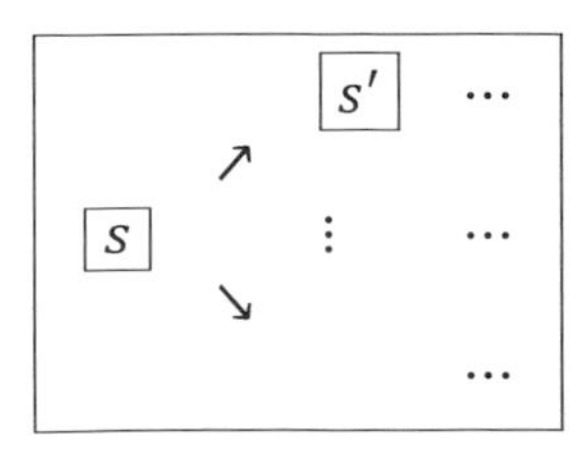

Let us return to the following question:

What is Peter: a lawyer or an economist?

This question can be formalized by the formula $?_i\{\alpha, \beta\}$. Its askability in a state $(\mathbf{M}, s)$ requires (at least) a minimal substructure on sR_i consisting of (at least) two accessible states, one of them satisfies α (and does not satisfy β) and the other one β (and does not satisfy α). All states in sR_i must satisfy the prospective presupposition $(\alpha \vee \beta)$, because of the context condition.

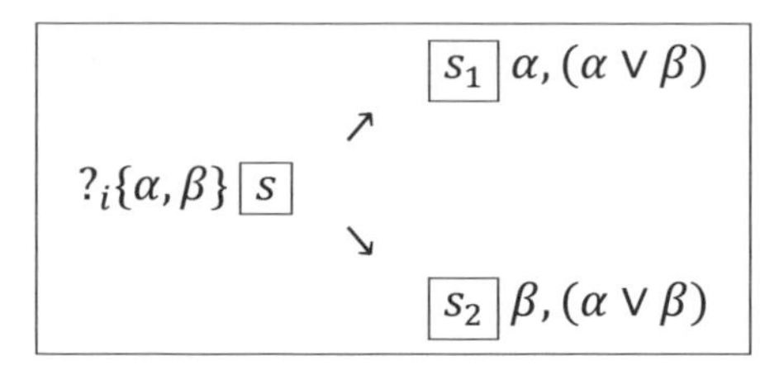

Of course, this is a minimal requirement given by askability conditions, the complete afterset structure can contain other states, some of them may satisfy both α and β, but none of them satisfies $\neg\alpha$ and $\neg\beta$; the question $?_i\{\alpha, \beta\}$ does not consider the answer *neither α nor β* as possible (context condition)—such an answer would be accepted, e.g., by the question

$$?_i\{\alpha, \beta, (\neg\alpha \wedge \neg\beta)\}$$

States in the afterset sR_i are understood as epistemic possibilities. In accordance with the non-triviality condition, neither α nor β can be true in all of them. Finally, the admissibility condition requires that there must be at least one 'α-state' and at least one 'β-state' in sR_i.

3.2.2 SOME IMPORTANT CLASSES OF QUESTIONS

In this section we introduce some classes of questions and their semantic behavior in our epistemic setting. Yes-no questions and conjunctive questions were introduced in the previous chapters. Hereafter, we suggest to link conditional and hypothetical questions with the role of context, where 'context' can be expressed by an auxiliary set of formulas. The names of the classes of questions we will use are borrowed from IEL (cf. [51]).

The very basic questions in their syntactical as well as semantical form are *yes-no questions.*

Is Prague the capital of the Czech Republic?

is a question requiring one of the following answers:

- (Yes,) Prague is the capital of the Czech Republic.
- (No,) Prague is not the capital of the Czech Republic.

with the formalization

$$?_i\{\alpha, \neg\alpha\},$$

which is written shortly as $?_i\alpha$. This question is askable in a state s if there are (at least) two different states available from s, one satisfies α and the other one $\neg\alpha$. The following substructure is required:

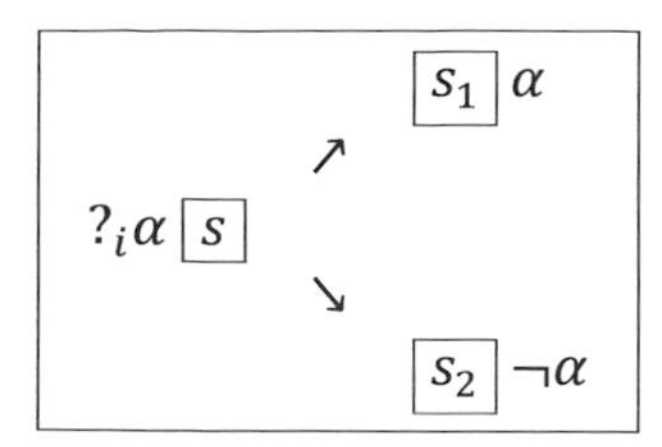

Reviewing the askability conditions for $?_i\alpha$ we can see that this question is equivalent to the formula $\langle i\rangle\alpha \wedge \langle i\rangle\neg\alpha$. In our system, yes-no questions can be seen as a 'contingency modality'. The same can be said about $?_i\neg\alpha$, both $?_i\alpha$ and $?_i\neg\alpha$ are equivalent. Yes-no questions always form a partitioning on aftersets and their presuppositions are tautologies. Questions with presuppositions, which are all tautological, are called *safe*.

Definition 21. *A question Q^i is* safe *iff* $\mathsf{PPres}Q^i$ *is valid.*

Questions that are not safe, will be called *risky*.[6]

Another example of safe questions are *conjunctive questions.* The shortest one is

$$?_i\{(\alpha \wedge \beta), (\neg\alpha \wedge \beta), (\alpha \wedge \neg\beta), (\neg\alpha \wedge \neg\beta)\}$$

asking for a full description based on α and β. We write it as $?_i|\alpha, \beta|$ and the following figure shows the required substructure of the afterset.

s_1 α, β

$?_i|\alpha, \beta|$ s $\nearrow$ $\longrightarrow$ $\searrow$ $\downarrow$

s_2 $\alpha, \neg\beta$

s_3 $\neg\alpha, \neg\beta$

s_4 $\neg\alpha, \beta$

Similarly to yes-no questions, they have an exhaustive set of direct answers, and direct answers are *mutually exclusive.* In Section 3.4 we will deal with some restrictions posed to direct answers which will enable us to shed more light on answerhood conditions.

The question

What is Peter: a lawyer or an economist?

with a formalization of $?_i\{\alpha, \beta\}$ is a risky one. Nonetheless, if the question is askable in a state, an agent does not admit that she has accessible states where $\mathsf{PPres}(?_i\{\alpha, \beta\})$ is not satisfied. Let us define this kind of 'local' safeness:

6 The original concepts of the safety and riskiness of questions come from Nuel Belnap. See Chapter 2 in this book as well.

Definition 22. *A question Q is* safe (for an agent i) in a state $(\mathbf{M}, s)$ *iff* $(\mathbf{M}, s_1) \vDash \mathsf{PPres}Q^i$, *for each* $s_1 \in sR_i$.

A question, which is askable in a state (for an agent), is safe in this state (for this agent).

Now we move to questions that will correspond to the idea of 'posing a question with respect to a set of auxiliary formulas'.

Conditional questions Let us consider the following question:

Did you stop smoking?

At first sight, it is a yes-no question, but seeing both answers it seems that there is something more that is presupposed:

- *Yes, I did* can mean *I had smoked and stopped.*
- *No, I didn't* can mean *I smoked and go on.*

Both of them presuppose the smoking in the past. Such a question is an example of a *conditional yes-no question* with a formalization $?_i\{(\alpha \wedge \beta), (\alpha \wedge \neg\beta)\}$. Generally, *conditional questions* are of the form

$$?_i\{\alpha \wedge \beta_1, \alpha \wedge \beta_2, \dots, \alpha \wedge \beta_n\}$$

The askability of a conditional question in a state s requires the validity of α in each accessible state, i.e., an agent 'knows' α in s. The concept of an askable (finite) conditional question can be generalized.

Definition 23. *A question Q is askable (by an agent i) in $(\mathbf{M}, s)$ with respect to a set of formulas Γ iff $(\mathbf{M}, s) \vDash \{[i]\gamma \mid \gamma \in \Gamma\}$ and $(\mathbf{M}, s) \vDash Q^i$.*

A conditional question $?_i\{\alpha \wedge \beta_1, \alpha \wedge \beta_2, \dots, \alpha \wedge \beta_n\}$ is askable in s if and only if $?_i\{\beta_1, \dots, \beta_n\}$ is askable there with respect to the auxiliary set of formulas (knowledge database) $\{\alpha\}$.

This leads us to the concept of *relativized askability*.[7] The term *relativized askability* will be used mostly for pointing out the importance of a (possibly big) auxiliary set of declaratives Γ. Clearly, every question askable in a state is askable with respect to its set of (prospective) presuppositions.

Fact 27. *If Q_i is askable in $(\mathbf{M}, s)$ with respect to Γ, then Q_i is askable in $(\mathbf{M}, s)$ with respect to every $\Delta \subseteq \Gamma$.*

However, relativized askability is not 'monotonic' in knowledge databases. If $(\mathbf{M}, s) \vDash Q_i$ with respect to a set Γ, then it need not be $(\mathbf{M}, s) \vDash Q_i$ with respect to a set $\Delta \supset \Gamma$.

Relativized askability of this kind can be useful for an explicit expressing of the knowledge structure. Catherine's question

---

7 This concept corresponds to the *question in an information set* introduced in [16].

Who has the Joker: Anne, or Bill?

can be understood as a question $?_c\{\alpha, \beta\}$ with respect to $(\neg\alpha \vee \neg\beta)$.

Hypothetical questions Inferential erotetic logic introduces one more term, *hypothetical question*, which is a bit similar to the previous one. A natural language example of a hypothetical yes-no question might be

If you open the door, will you see a bedroom?

with a formalization $?_i\{(\alpha \to \beta), (\alpha \to \neg\beta)\}$. A general hypothetical question is then

$$?_i\{\alpha \to \beta_1, \dots, \alpha \to \beta_n\}$$

Again, the askability of such questions can be understood as based on an agent's hypothetical knowledge. Our interpretation is: if α is known, then it is to be decided whether β_1, or β_2, etc. Using a generalization similar to Definition 23 we obtain

Definition 24. *A question Q is askable (by an agent i) in $(\mathbf{M}, s)$ with respect to a set of hypotheses Γ iff $(\mathbf{M}, s) \models \{[i]\gamma \mid \gamma \in \Gamma\}$ implies $(\mathbf{M}, s) \models Q^i$.*

The askability of a conditional question ensures the askability of a hypothetical one. The difference in the formulation of both definitions lies in the words *and* and *implies*. We will return to these concepts in Section 3.5.

3.3 EPISTEMIC EROTETIC IMPLICATION

Erotetic inference is implicitly based on (standard) implication. We say that a question Q_1 implies Q_2 (in a state s, for an agent i) whenever askability of Q_1 (in s, for i) implies askability of Q_2 (in s, for i).

$$(\mathbf{M}, s) \models Q_1^i \to Q_2^i \text{ iff } (\mathbf{M}, s) \models Q_1^i \text{ implies } (\mathbf{M}, s) \models Q_2^i$$

We have mentioned that questions $?_i\alpha$ and $?_i\neg\alpha$ have the same askability conditions. The eqivalence of both questions is a theorem in our system based on modal logic K. Let us omit the index i for now.

Example 14. *Both $(?\alpha \to ?\neg\alpha)$ and $(?\neg\alpha \to ?\alpha)$ are valid.*

The informal meaning of epistemic erotetic implication is very transparent. Whenever an agent asks Q, then she can ask every question implied by Q. A question in the antecedent is 'stronger' (in the sense of inference) than the implied one. The afterset substructure required by an implied question must be a substructure of that required by an implying question. The question

What is Peter: a lawyer or an economist?

implies

Is Peter a lawyer?

as well as

Is Peter an economist?

This can be generalized:

Example 15. $?\{\alpha_1, \dots, \alpha_n\} \rightarrow ?\alpha_j$ *is valid, for each* $j \in \{1, \dots, n\}$.

The following example shows the special position of conjunctive questions in implications.

Example 16. *The following implications are valid:*

1. $?|\alpha_1, \dots, \alpha_n| \rightarrow ?\alpha_j$*, for each* $j \in \{1, \dots, n\}$.
2. $?|\alpha, \beta| \rightarrow ?(\alpha \circ \beta)$*, where* $\circ$ *is any truth-functional constant.*
3. $?|\alpha, \beta| \rightarrow ?\{\alpha, \beta, (\neg\alpha \wedge \neg\beta)\}$
4. $?|\alpha, \beta, \gamma| \rightarrow ?|\alpha, \beta|$[8]

Let us notice that conjunctive questions are safe and they imply safe questions in the example. We can prove that the safeness in a state is transferred by implication.

Proposition 24. *If* Q_1 *is a safe question (in the state* $(\mathbf{M}, s)$*) and the formula* $(Q_1 \rightarrow Q_2)$ *is valid, then* Q_2 *is safe in the state* $(\mathbf{M}, s)$.

Proof. Let us suppose that Q_1 is safe and $(Q_1 \rightarrow Q_2)$ is valid. If Q_1 is safe (Definition 21), then it is safe in any epistemic state (Definition 22). Thus, for any model $\mathbf{M}$ and state s, if $(\mathbf{M}, s) \models Q_1$, then $(\mathbf{M}, s) \models Q_2$, because of the validity of $(Q_1 \rightarrow Q_2)$. Since Q_2 is askable in $(\mathbf{M}, s)$, it is also safe in $(\mathbf{M}, s)$ (condition 3 in Definition 20). □

It is easy to check that neither $?|\alpha, \beta| \rightarrow ?\{\alpha, \beta\}$ nor $?\{\alpha, \beta\} \rightarrow ?|\alpha, \beta|$ are valid. In both cases there is a problem with the context condition; a risky question $?\{\alpha, \beta\}$ requires the validity of the disjunction $(\alpha \vee \beta)$ in the afterset. The next two examples emphasize the importance of the context condition again.

Example 17. $\not\models ?\{\alpha, \beta\} \rightarrow ?\{\neg\alpha, \neg\beta\}$ *as well as* $\not\models ?\{\neg\alpha, \neg\beta\} \rightarrow ?\{\alpha, \beta\}$

Example 18. $\not\models ?\{\alpha, \beta, \gamma\} \rightarrow ?\{\alpha, \beta\}$

---

8 Mariusz Urbański pointed out the following generalization: $?|\alpha_1, \dots, \alpha_n|$ implies each question $?|\beta_1, \dots, \beta_j|$ where $\{\beta_1, \dots, \beta_j\} \subseteq \{\alpha_1, \dots, \alpha_n\}$.

The question $?_i\{\alpha, \beta, \gamma\}$ requires $[i](\alpha \vee \beta \vee \gamma)$, but $?_i\{\alpha, \beta\}$ requires 'only' $[i](\alpha \vee \beta)$, which can fail in the structure sufficient for the askability of the first question.

An implying question shares presuppositions with the implied one.

Fact 28. *If* $Q_1 \rightarrow Q_2$ *is valid, then if* $\varphi \in \mathsf{Pres}Q_1$*, then* $\varphi \in \mathsf{Pres}Q_2$.

Epistemic erotetic implication has the expected property—transitivity:

Fact 29. *If* $(\mathbf{M}, s) \models Q_1 \rightarrow Q_2$ *and* $(\mathbf{M}, s) \models Q_2 \rightarrow Q_3$*, then* $(\mathbf{M}, s) \models Q_1 \rightarrow Q_3$.

3.4 ASKABILITY AND ANSWERHOOD

Epistemic erotetic implication creates a relationship among questions. If a question is askable in a state, so is every implied one. In our system based on classical logic, $Q_1 \rightarrow Q_2$ if and only if $\neg Q_2 \rightarrow \neg Q_1$. If an implied question is non-askable, so is the implying one. The askability of a question consists of the validity of three conditions (non-triviality, admissibility, and context) and non-askability is a result of a violation of at least one of them. Let us imagine that we know that $Q_1 \rightarrow Q_2$ and we have an answer to Q_2 (non-triviality condition fails), then we are sure that Q_1 is non-askable, but does it mean it has an answer to Q_1? What if Q_1 is non-askable because the context condition fails? In this section we will just deal with such violations of askability conditions, properties of non-askable questions from various classes of questions, and answerhood conditions—complete and partial answers will be introduced.

To break the non-triviality condition means that there is a direct answer which is 'known' by an agent (in a state of a model). In fact, an agent knows a direct answer even if she knows a formula that is equivalent to a direct answer or she knows a formula from which some direct answer follows. Such a formula is called a *complete answer.*

Let us define the concept of an *answered question* in a state (for an agent):[9]

Definition 25. *A question* $Q^i = ?_i\{\alpha_1, \dots, \alpha_n\}$ *is* answered *in* $(\mathbf{M}, s)$ *(for an agent i) iff* $(\mathbf{M}, s) \models \bigvee_{\alpha_j \in dQ^i} ([i]\alpha_j)$. *We write* $(\mathbf{M}, s) \models A_i Q$.

The case of an invalid admissibility condition is a bit different. Let us imagine that an agent knows α in a state s. Then, even if she does not know an answer to a question $?|\alpha, \beta|$ in s, it is not right to ask this question. Not all possibilities required by the admissibility condition are available, in particular, accessible states with $\{\neg\alpha, \neg\beta\}$ and $\{\neg\alpha, \beta\}$ are missing. Some answers to the question

9 We do not discuss a dynamic approach just now. An answered question is a 'potentially' answered question, in fact, an answer need not be uttered among agents.

$?|\alpha, \beta|$ give information that is superfluous to the state of the agent's knowledge. Formula α is a *partial answer* to a question $?|\alpha, \beta|$. A partial answer excludes some of the (direct) answers.

Definition 26. *A question* $Q^i = ?_i\{\alpha_1, \dots, \alpha_n\}$ *is* partially answered *in* $(\mathbf{M}, s)$ *(for an agent i) iff* $(\mathbf{M}, s) \vDash \bigvee_{\alpha_j \in dQ^i} ([i]\neg\alpha_j)$. *We write* $(\mathbf{M}, s) \vDash P_i Q$.

In fact, if the admissibility condition fails, there is a direct answer which is not considered as possible, i.e., there is $\alpha \in dQ^i$ such that $(\mathbf{M}, s) \nvDash \langle i \rangle \alpha$. Equivalently, in our system, $(\mathbf{M}, s) \vDash [i]\neg\alpha$. This means that our agent can answer the question $?\alpha$ in $(\mathbf{M}, s)$: $(\mathbf{M}, s) \vDash A_i?\alpha$ for some $\alpha \in dQ^i$. Together with Example 15 we have just obtained the proof of the following proposition:

Proposition 25. *If* $(\mathbf{M}, s) \vDash P_i Q$, *then there is a formula* φ *such that* $(\mathbf{M}, s) \vDash A_i?\varphi$ *and* $Q^i \to ?_i\varphi$ *is valid.*

Let us suppose, a question Q is answered. Does it mean that Q is partially answered? Surprisingly not. It is not true that if $(\mathbf{M}, s) \vDash A_i Q$, then $(\mathbf{M}, s) \vDash P_i Q$. See the next example.

Example 19.

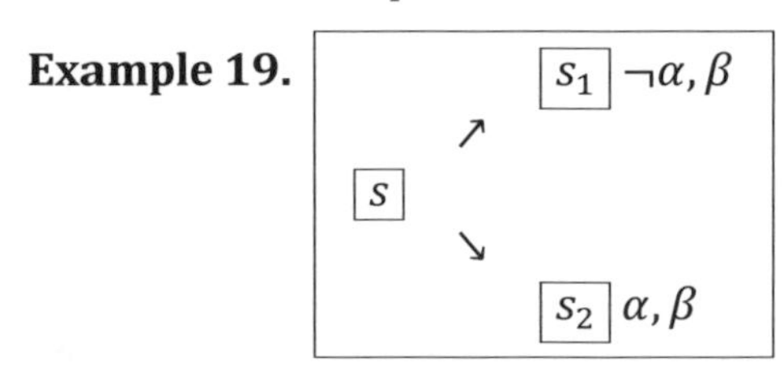

In the structure of Example 19 the question $?\{\alpha, \beta\}$ is answered in s (the agent knows β), but it is not partially answered; the agent is not able to get the knowledge of either $\neg\alpha$ or $\neg\beta$, both α and β are still possible.

$A_i Q$ implies $P_i Q$ for questions with *pairs of mutually exclusive* direct answers. It means that for each direct answer there is another one such that both of them cannot be true. Yes-no questions as well as conjunctive questions are examples from this class. Their sets of direct answers satisfy a more strict condition, they have *mutually exclusive* direct answers—the 'truth' of a direct answer (in a state) means that no other direct answer is satisfied there. Both kinds of mutual exclusiveness of direct answers are of a semantic nature, they can be caused by a state in a model. Recall Catherine's question

Who has the Joker: Anne, or Bill?

with the context, which is important here: the context does not admit the after-set substructure in Example 19.

Mutual exclusiveness is not preserved by implication. We cannot prove anything similar to Proposition 24. In particular, the answers to the question $?_i\{\alpha \wedge \beta, \alpha \wedge \neg\beta, \neg\alpha \wedge \beta\}$ are mutually exclusive, but this question implies $?_i\{\alpha, \beta\}$, which need not be in this class, generally.

Fact 30. *If dQ^i consists of pairs of mutually exclusive formulas, then A_iQ implies P_iQ.*

The last option for non-askability is the violation of the context condition. There, an agent does not know (believe) the prospective presupposition of a question. For example, the question

Which town is the capital of the Czech Republic: Prague, or Brno?

can be non-askable, although an agent does not know either complete or partial answer, but she admits that there is another town, which could be the capital of the Czech Republic, i.e., neither Prague nor Brno might be the right answer. Then we say that a question is *weakly presupposed* by an agent.

Definition 27. *A question Q^i is* weakly presupposed *in $(\mathbf{M}, s)$ (by an agent i) iff $(\mathbf{M}, s) \not\models [i]\mathsf{PPres}Q^i$. We write $(\mathbf{M}, s) \models W_iQ$.*

An askable question in a state (for an agent i) satisfies at least the 'safeness in a state' (see Definition 22). If a question Q is safe in $(\mathbf{M}, s)$, then Q cannot be weakly presupposed in $(\mathbf{M}, s)$.

Proposition 26. *If a question Q^i is safe in $(\mathbf{M}, s)$ and dQ^i consists of pairs of mutually exclusive formulas, then the following conditions are equivalent:*

1. $(\mathbf{M}, s) \models \neg Q^i$
2. $(\mathbf{M}, s) \models P_iQ$
3. *There is a formula φ such that $(\mathbf{M}, s) \models A_i?\varphi$ and $Q^i \rightarrow ?_i\varphi$ is valid.*

Proof. (2⇒1) is clear and (2⇒3) is from Proposition 25.

(1⇒2) If $(\mathbf{M}, s) \models \neg Q^i$, then there are three possibilities: $(\mathbf{M}, s) \models A_iQ$ or $(\mathbf{M}, s) \models P_iQ$ or $(\mathbf{M}, s) \models W_iQ$. The last one is impossible because of the safeness of Q^i in $(\mathbf{M}, s)$. If $(\mathbf{M}, s) \models A_iQ$, then $(\mathbf{M}, s) \models P_iQ$ (from Fact 30).

(3⇒1) Let us suppose that there is a formula φ such that the question $?_i\varphi$ is answered in $(\mathbf{M}, s)$. From $Q^i \rightarrow ?_i\varphi$ we know that if $(\mathbf{M}, s) \not\models ?_i\varphi$, then $(\mathbf{M}, s) \not\models Q^i$. □

Partial answerhood of a question Q^i in some state is equivalent to the existence of a yes-no question, which is answered in that state and implied by Q^i. From the validity of $Q^i \rightarrow ?_i\varphi$ we know that the non-askability of $?_i\varphi$[10] implies non-askability of Q^i and, therefore, φ (as well as $\neg\varphi$) implies either some $\alpha \in dQ^i$ or $\neg\alpha$, for $\alpha \in dQ^i$.

---

10 $(\mathbf{M}, s) \models \neg ?_i\varphi$ iff $(\mathbf{M}, s) \models A_i?\varphi$.

3.5 CONTEXT

In Section 3.2.2 we introduced the concept of askability with respect to auxiliary sets of formulas. Although both introduced types are understood as variants of conditional or hypothetical questions, it can be useful to display and emphasize the role of context in some situations. In particular, if it has an important position in reasoning. Simultaneously, questions are required to have finite sets of direct answers.

Let us recall the example from Section 2.3.1, where we discussed erotetic implication in IEL. An agent asking

> Q_1: *Which faculty did Peter graduate from: a faculty of law or a faculty of economics?*

can be satisfied by the answer

> *He is a lawyer.*

even if she did not ask

> Q_2: *What is Peter: a lawyer or an economist?*

The connection between both questions could be established by the following knowledge base Γ:

> *Someone is a graduate of a faculty of law iff he/she is a lawyer.*
> *Someone is a graduate of a faculty of economics iff he/she is an economist.*

Relativized askability helps us to express that Q_1 implies Q_2 with respect to an auxiliary set of formulas Γ. In the example, Q_1 can be formalized by $?\{\alpha_1, \alpha_2\}$, Q_2 by $?\{\beta_1, \beta_2\}$, and the background knowledge is $\Gamma = \{(\beta_1 \leftrightarrow \alpha_1), (\beta_2 \leftrightarrow \alpha_2)\}$.

As we have just said, the primary reason for introducing askability with respect to an auxiliary set of formulas is to emphasize the importance of context in inferences with questions. This can be considered a generalization of conditional and hypothetical questions in our system. Conditional questions consist of two parts: a conditional part (context) and a query part. As an easy conclusion to Fact 27, we receive that a conditional question implies its query part:

$$?_i\{\alpha \wedge \beta_1, \dots, \alpha \wedge \beta_n\} \rightarrow ?_i\{\beta_1, \dots, \beta_n\}.$$

In the generalized version we have

Fact 31. *If Q^i is askable in $(\mathbf{M}, s)$ with respect to Γ, then $(\mathbf{M}, s) \vDash Q^i$.*

This is not true for hypothetical questions. $?_i\{\alpha \rightarrow \beta_1, \dots, \alpha \rightarrow \beta_n\}$ does not imply $?_i\{\beta_1, \dots, \beta_n\}$.

In some sense, we can see (generalized) hypothetical questions as a counterpart of evocation in IEL. Similarly, (generalized) conditional questions in the interplay with implication seem to be a counterpart of general erotetic implication in IEL. The correspondence could be seen in a similar 'philosophy', but these concepts differ in properties.

To be able to speak about the properties of these concepts we have to introduce some abbreviations. First, let us write $(\mathbf{M}, s) \models [i]\Gamma$ instead of $(\mathbf{M}, s) \models \{[i]\gamma \mid \gamma \in \Gamma\}$. Second, whenever we want to express that a question Q_1^i is askable by an agent i in $(\mathbf{M}, s)$ with respect to a set of formulas Γ (Definition 23), we will write

$$(\mathbf{M}, s) \models ([i]\Gamma \bowtie Q_1^i)$$

If, moreover, Q_2^i is implied (by Q_1^i with respect to Γ) in this state, we write

$$(\mathbf{M}, s) \models ([i]\Gamma \bowtie Q_1^i) \Rightarrow Q_2^i$$

The last abbreviation we are going to introduce expresses that a question Q is askable in a state $(\mathbf{M}, s)$ with respect to a set of hypotheses Γ (Definition 24):

$$(\mathbf{M}, s) \models [i]\Gamma \Rightarrow Q^i$$

Now we can list some properties appearing in combinations of (generalized) conditional and hypothetical questions with implication in an readable manner. Most of them are apparent and expected. For example, the following fact points out the cumulativity of explicitly expressed presuppositions:

Fact 32. *If* $(\mathbf{M}, s) \models ([i]\Gamma \bowtie Q_1^i) \Rightarrow Q_2^i$ *and* $(\mathbf{M}, s) \models ([i]\Delta \bowtie Q_2^i) \Rightarrow Q_3^i$, *then* $(\mathbf{M}, s) \models ([i](\Gamma \cup \Delta) \bowtie Q_1^i) \Rightarrow Q_3^i$.

Relativized askability is transferred by implication:

Fact 33. *If* $(\mathbf{M}, s) \models ([i]\Gamma \bowtie Q_1^i)$ *and* $(\mathbf{M}, s) \models (Q_1^i \to Q_2^i)$, *then* $(\mathbf{M}, s) \models ([i]\Gamma \bowtie Q_2^i)$.

It is easy to check the following generalization:

Proposition 27. *If* $(\mathbf{M}, s) \models ([i]\Gamma \bowtie Q_1^i)$ *and* $(\mathbf{M}, s) \models ([i]\Gamma \bowtie Q_1^i) \Rightarrow Q_2^i$, *then* $(\mathbf{M}, s) \models ([i]\Gamma \bowtie Q_2^i)$.

The same result can be proved for generalized hypothetical questions:

Fact 34. *If* $(\mathbf{M}, s) \models [i]\Gamma \Rightarrow Q_1^i$ *and* $(\mathbf{M}, s) \models (Q_1^i \to Q_2^i)$, *then* $(\mathbf{M}, s) \models [i]\Gamma \Rightarrow Q_2^i$.

Finally, we have a generalization of the previous fact:

Proposition 28. *If* $(\mathbf{M}, s) \models [i]\Gamma \Rightarrow Q_1^i$ *and* $(\mathbf{M}, s) \models ([i]\Gamma \bowtie Q_1^i) \Rightarrow Q_2^i$, *then* $(\mathbf{M}, s) \models [i]\Gamma \Rightarrow Q_2^i$.

Proof. Let us suppose $(\mathbf{M}, s) \nvDash [i]\Gamma \Rightarrow Q_2^i$, i.e., $(\mathbf{M}, s) \vDash [i]\Gamma$ and $(\mathbf{M}, s) \nvDash Q_2^i$. From $(\mathbf{M}, s) \vDash [i]\Gamma$ and $(\mathbf{M}, s) \vDash [i]\Gamma \Rightarrow Q_1^i$ we get $(\mathbf{M}, s) \vDash Q_1^i$ and $(\mathbf{M}, s) \vDash ([i]\Gamma \bowtie Q_1^i)$. Because of $(\mathbf{M}, s) \vDash ([i]\Gamma \bowtie Q_1^i) \Rightarrow Q_2^i$ we obtain $(\mathbf{M}, s) \vDash Q_2^i$, which is a contradiction. □

It is necessary to remark once more that both variants of relativized askability were introduced for the explicit expression of the context conditions of a question. This approach is useful; however, (pure) questions and implication will be of primary importance in our finite communication setting in Chapter 4.

3.6 IMPLIED QUESTIONS

If questions are in the implicational relationship, the transmission of askability conditions from an implying question to the implied one is justified. An implied question is understood as 'less complex' in its requirements posed on the afterset substructure. Let us recall Example 15[11] where $Q \rightarrow ?\alpha$ is valid for each $\alpha \in dQ$, thus

$$Q \rightarrow \bigwedge_{\alpha \in dQ} (?\alpha)$$

On the one hand, each question $?\alpha$ is a yes-no question with many good properties. On the other hand, they may be 'worse' in the description of an agent's knowledge/ignorance structure than the initial question Q. Might it be 'better' if we consider the whole set of implied questions based on the set of direct answers to an initial question?

If we consider the set of questions $\Phi = \{?\alpha_1, ?\alpha_2, ...\}$, then the non-askability of some $?\alpha$ means that the answer is either α or $\neg\alpha$. In the first case Q is answered as well, in the second one Q is partially answered. This can be understood as a form of 'sufficiency' condition of the set Φ: the answerability of its members implies at least the partial answerability of the initial question Q. It means Φ does not include 'useless' questions. Moreover, we receive one more property, Φ is 'complete' in a way: if Q is partially answered, then there must be a question $Q_j \in \Phi$ that is answered. We can understand it as a form of 'reducibility' of an initial question to a set of yes-no questions. In comparison with IEL, this reducibility is based purely on implication.

Example 16 gives a similar result for conjunctive questions. The set of yes-no questions can be formed by constituents of their direct answers:

$$?|\alpha_1, \dots, \alpha_n| \rightarrow \bigwedge_{j=1}^{n} (?\alpha_j)$$

11 Let us omit the index i in this section.

In this case, we arrive at really less complex questions, but having a partial answer to $?|\alpha_1, \dots, \alpha_n|$ does not give either an answer or a partial answer to some α_j. It could be useful in some cases. An agent can ask questions from the set $\Phi = \{?\alpha_1, \dots, ?\alpha_n\}$ and complete her knowledge step by step. The most important property is that the set Φ does not include useless questions. Generally speaking, in some communication processes it is useful to conceal some knowledge or ignorance of a questioner—a criminal investigation is a nice example. An agent can ask questions from the set Φ without completely revealing her knowledge structure. Asking a conjunctive question $?|\alpha_1, \dots, \alpha_n|$ publicly, everybody is informed that the agent-questioner does not know anything with respect to $\alpha_1, \dots, \alpha_n$.

4. A STEP TOWARDS THE DYNAMIZATION OF EROTETIC LOGIC

4.1 INTRODUCTION

At the beginning of the previous chapter we introduced the language for multi-agent propositional epistemic logic $\mathcal{L}^K_{cpl}$ with a finite set of agents $\mathcal{A} = \{1, \dots, m\}$. If we add an accessibility relation for each agent to Kripke frames, $\mathcal{F} = \langle S, R_1, \dots, R_m \rangle$, we will obtain the multi-modal system K with box-like modalities $[1]$, ..., $[m]$ in the role of the agents' knowledge. There are many discussions about the best representation of knowledge, as well as a belief in the philosophy of logic. Nowadays, such discussions are brought to life again in the studies of substructural logics. The term *knowledge* is often subjected to new interpretations based on a background system.[1]

The very aim of a question's implementation in epistemic-like systems was to provide an interpretation of questions, which corresponds with the interplay of the idea of representing the knowledge and ignorance structure of a questioner in the process of asking and answering. The interpretation of questions should be, ideally, independent of a background system. However, in our philosophy, a 'knowledge structure' and its representation is considered to be primary. We follow the mainstream in epistemic logic and work with the multi-modal system S5. The step toward dynamization will be made in the dynamic extension of this system called *public announcement logic*. S5 represents standard epistemic logic, cf. [48, 11], where knowledge is factive and fully introspective (positively

1 See, e.g., the definition of knowledge modality inside relevant logic in [2].

as well as negatively). On the one hand, this system is subjected to criticism, see [11, 8] where the 'logical omniscience problem' seems to be the most criticized aspect. On the other hand, the goal of this chapter is to show the role of questions in a formal dynamic-epistemic system. So, we are not going to solve any problem of this formal epistemic representation nor follow with philosophical discussions on it.

On the previous pages, we often referred to the importance of questions in the communication processes. Communication is understood as an information exchange among agents in a group. *Public announcement logic* is one of the models of communication among agents. The delivering of information in a group influences a change of the epistemic states of group members.

First of all we extend the erotetic epistemic framework by group questions and group epistemic modalities (group knowledge, common knowledge, and distributed knowledge). This makes it possible to speak about answerhood conditions for groups of agents. Then we apply the public announcement modality in the process of answer mining among agents.

4.2 MULTI-AGENT PROPOSITIONAL EPISTEMIC LOGIC WITH QUESTIONS

We have just said that our epistemic framework is based on multi-modal logic S5.[2] The accessibility relations $R_1, \dots, R_m$ are *equivalence* relations, i.e., reflexive, transitive, and symmetric. In logic K we understand the accessible states as (epistemic) 'alternatives' for an actual state, an agent can see epistemic 'possibilities'. Accessibility relations in S5 seem to play a bit different role. The equivalence relation between two epistemic states makes them 'indistinguishable', an agent considers such epistemic states as having the same epistemic 'worth'.

Let us recall the group of three friends and just one Joker distributed among them. From Catherine's viewpoint both possibilities—either *Anne has the Joker (and Bill has not)* or *Bill has the Joker (and Anne has not)*—are indistinguishable:[3]

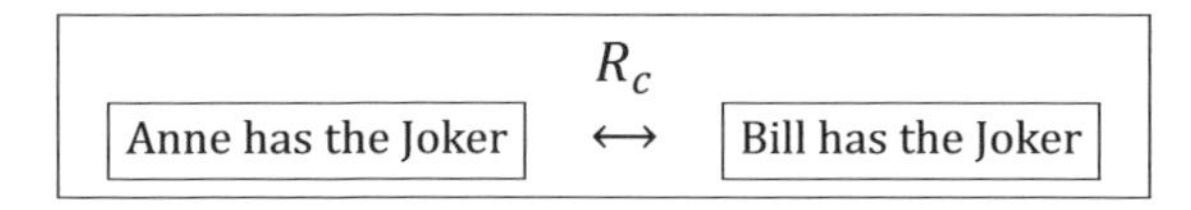

2 Axiomatics is given in the last section 4.4.

3 We omit the reflexive loops in all figures in this section.

4.2.1 GROUP EPISTEMIC MODALITIES

So far we have worked with individual knowledge. In multi-agent systems we have to introduce new modalities to reflect epistemic states in groups of agents. The language $\mathcal{L}^K_{cpl}$ will be extended by the symbols E_G, C_G, and D_G, where $G \subseteq \mathcal{A}$ is a group of agents.

Group knowledge Formula $E_G\varphi$ is interpreted as

'each agent (from G) knows φ'

and is fully definable by individual knowledge operators of all G-members:

$$E_G\varphi \leftrightarrow \bigwedge_{i \in G} [i]\varphi$$

The *group knowledge* of φ does not guarantee that one member of the group G knows that another one has the same knowledge about φ. The agents in G need not share any information with other members of the group. Group knowledge is not sufficient for coordination of common activities in a group.

Common knowledge The group modality C_G interprets the conditions for sharing of information. Formula $C_G\varphi$ expresses not only that φ is group knowledge, but also that this fact is reflected by everybody in G:

> Each agent (from G) knows φ and each agent knows that each agent knows φ and each agent knows that each agent knows that each agent knows φ and ...

Common knowledge of a fact means that the knowledge of it is shared by everybody in G and, moreover, each agent is aware of this sharing. C_G can be seen as an infinite conjunction of all finite iterations of the group knowledge E_G:

$$C_G\varphi \leftrightarrow E_G\varphi \wedge E_G E_G\varphi \wedge E_G E_G E_G\varphi \wedge \ldots$$

The system S5C is obtained by adding the operator C_G and the semantic clause:

- $(\mathbf{M}, s) \models C_G\varphi$ iff $(\mathbf{M}, t) \models \varphi$ for each t such that $s \left(\bigcup_{i \in G} R_i\right)^* t$

The relation $\left(\bigcup_{i \in G} R_i\right)^*$ is a reflexive and transitive closure of $\bigcup_{i \in G} R_i$. It means here that there are states $s_1, \ldots, s_k$ such that $s_1 = s$ and $s_k = t$ and for all j with $1 \leq j \leq k$ there exists $i \in G$ such that $s_j R_i s_{j+1}$.

As we said, E_G is definable in the language $\mathcal{L}^K_{cpl}$, so adding group knowledge is just a conservative extension of the background multimodal epistemic logic S5. Both languages $\mathcal{L}^K_{cpl}$ and $\mathcal{L}^{KE}_{cpl}$ have the same expressivity. However, this is not the case with common knowledge. Multi-modal epistemic logic with common knowledge S5C is not compact, as it is indicated in the definition of C_G, and there exists a formula in the language $\mathcal{L}^{KC}_{cpl}$ which can distinguish between two models that are indistinguishable with respect to language $\mathcal{L}^{K}_{cpl}$, see [48, p. 227].

The relationship of the introduced epistemic modalities is the following:

Fact 35. *$C_G\varphi \to E_G\varphi \to [i]\varphi$ is valid in* S5C *for each $i \in G$.*

Common knowledge is essential for collective behavior and the coordination of collective actions. In game theory, it is often presupposed that the rules of a game are shared by its players. It is important to know the rules, to know that the other players know the same rules, to know that they know that we (players) know it, and so on. Common knowledge alone is a good candidate for group answerhood conditions. We say that a question is (partially) answered for a group of agents whenever a (partial) answer is commonly known.

Definition 28.

- *A question Q is* answered *in $(\mathbf{M}, s)$ for group G iff there is $\alpha \in dQ$ such that $(\mathbf{M}, s) \vDash C_G\alpha$.*
- *A question Q is* partially answered *in $(\mathbf{M}, s)$ for group G iff there is $\alpha \in dQ$ such that $(\mathbf{M}, s) \vDash C_G(\neg\alpha)$.*

Distributed knowledge This group epistemic modality is a bit of another kind. Let us recall the group of three friends and suppose that Anne has the Joker. Although neither Catherine nor Bill knows it, the knowledge of the Joker-owner is implicitly contained in the group. If the agents (at least Bill and Catherine) can communicate, they easily reach the hidden fact that Anne has the card. The standard meaning of $D_G\varphi$ is given by the semantic clause:

- $(\mathbf{M}, s) \vDash D_G\varphi$ iff $(\mathbf{M}, t) \vDash \varphi$ for each t such that $s\left(\bigcap_{i \in G} R_i\right) t$

The formula φ, which is 'implicitly' known in group G, is true in all states that are accessible for each member of G. The operator D_G is sometimes called *distributed* knowledge and sometimes *implicit* knowledge. The term *distributed knowledge* coincides better with the idea of pooling agents' knowledge together. Let us imagine that a solution of some problem can be obtained by the collection of particular data from each member of a group of agents. The crucial data are 'distributed' among agents, but nobody can solve the problem alone because of the need for the other data.

If an agent knows φ, then φ is distributed knowledge in all groups, in which she is a member:

Fact 36. *$[i]\varphi \to D_G\varphi$ is valid in* S5CD *for each $i \in G$.*

The accessibility relation based on D_G is a subset of each R_i. Adding D_G to language $\mathcal{L}^K_{cpl}$ does not increase its expressivity.[4]

If there is distributed knowledge for a group of agents, then it is distributed knowledge for every bigger group.

Fact 37. *$D_G\varphi \to D_{G'}\varphi$ is valid in* S5CD *for $G \subseteq G'$.*

The role of distributed knowledge in (group) answerhood conditions will be discussed in the next section.

4.2.2 GROUP QUESTIONS AND ANSWERHOOD

We introduced questions in the form of an agent's personal task that fits in her knowledge structure. Whenever she wants to find an answer to a question, she has to communicate the question and find 'someone' who can answer it. Being in a group of colleagues she asks the question and, in the best case, there is somebody who knows the answer. The worst case is that nobody can answer the question—the question is the task for all of them. Such a question is askable by each member of group G and we shall call it *group question.*

Definition 29 (Group question). *A question Q is an* askable group question *in $(\mathbf{M}, s)$ (for a group of agents G) iff $(\forall i \in G)((\mathbf{M}, s) \models Q^i)$. Let us write $(\mathbf{M}, s) \models Q^G$.*

Whenever there is a question which is (partially) answered by at least one agent in a group, then we can see how to reach a (partial) answer. The question must be publicly posed and the received answer is a result of this communication.

Group questions seem to be a worse problem, there is no agent with a (partial) answer to it in a group. If an answer should be sought inside the group, there is only one chance to find it. Again, communication is important for to discover 'hidden' information. An answer can be present in the group as *distributed knowledge.*

Let us have a group of two agents a (Anne) and b (Bill). The following example shows their knowledge structure:

Example 20.

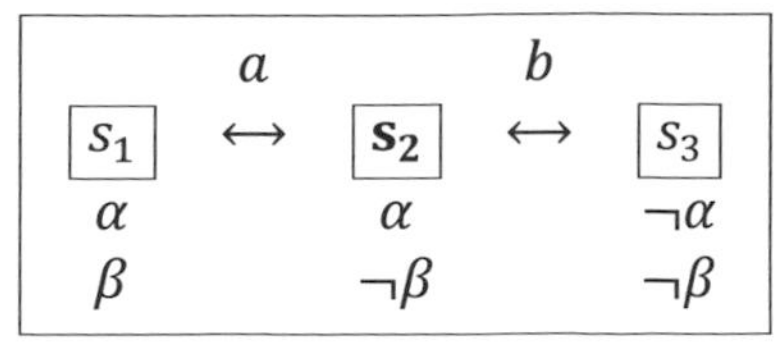

4 Axioms and properties can be found in [27].

Anne cannot distinguish between states s_1 and s_2 and she knows α, Bill cannot distinguish between states s_2 and s_3 and he knows $\neg\beta$. However, neither of them is able to (partially) answer the yes-no question $?(\alpha \to \beta)$, it is their group question $?_{\{a,b\}}(\alpha \to \beta)$. If they communicate, they recognize the state s_2 to be common for them. This brings us to the term *implicitly (partially) answered question.*

Definition 30.

- *A question Q is* implicitly answered *in* $(\mathbf{M}, s)$ *by a group of agents G iff* $(\exists\alpha \in dQ)((\mathbf{M}, s) \vDash D_G\alpha)$.
- *A question Q is* implicitly partially answered *in* $(\mathbf{M}, s)$ *by a group of agents G iff* $(\exists\alpha \in dQ)((\mathbf{M}, s) \vDash D_G\neg\alpha)$.

Back to the example, what should agents a and b communicate to obtain an answer to $?(\alpha \to \beta)$? We can find inspiration in implied yes-no questions. The question $?(\alpha \to \beta)$ implies the disjunction of questions $?\alpha$ and $?\beta$:

$$?(\alpha \to \beta) \to (?\alpha \lor ?\beta)$$

Above that, the agent a can completely answer $?\alpha$ and the other one can completely answer the question $?\beta$. So, it would be useful to communicate questions $?\alpha$ and $?\beta$ in the group.

Generally, we can prove that if there is a set of questions, their disjunction is implied by an initial question, and each question from the set can be (partially) answered by some agent from group G, then the initial question is implicitly (partially) answered.

Proposition 29. *If there is a set of questions Φ such that each $Q_k \in \Phi$ is (partially) answered in $(\mathbf{M}, s)$ by some agent $i \in G$ and $(\mathbf{M}, s) \vDash Q \to \bigvee_{Q_k \in \Phi} Q_k$, then Q is implicitly (partially) answered in $(\mathbf{M}, s)$.*

Proof. If Q is not an askable group question, then $\Phi = \{Q\}$ and there is an agent having a (partial) answer to Q. From Facts 35 and 36 we get that it is an implicit (partial) answer in group G.

Let us suppose Q is an askable group question in $(\mathbf{M}, s)$, then (partial) answers to questions from Φ are distributed among agents from G. From $(\mathbf{M}, s) \vDash Q \to \bigvee_{Q_k \in \Phi} Q_k$ we have $(\mathbf{M}, s) \vDash \bigwedge_{Q_k \in \Phi} \neg Q_k \to \neg Q$. The non-askability of Q in $(\mathbf{M}, s)$ cannot be caused by a violation of the context condition because of its status as an askable group question. Now, let us introduce a new (fictive) agent a, which pools knowledge of all agents in group G together: $R_a = \bigcap_{i \in G} R_i$. If there are no common epistemic states, $R_a = \emptyset$, then every formula is distributed knowledge. If R_a is nonempty, then all questions $Q_k \in \Phi$ are non-askable for a in $(\mathbf{M}, s)$, so is Q, and a's knowledge of a (partial) answer is in afterset sR_a. □

The proposition corresponds with the idea behind the following S5CD-valid rule

$$\frac{(\psi_1 \wedge \dots \wedge \psi_m) \to \varphi}{([l_1]\psi_1 \wedge \dots \wedge [l_m]\psi_m) \to D_{\{l_1,\dots,l_m\}}\varphi} \tag{4.1}$$

A formula φ can be logically obtained from the formulas $\psi_1, \dots, \psi_m$. If the knowledge of ψ-formulas is scattered among agents in a group (some agents can know more than one ψ-formula, some can know none of them), then φ is distributed knowledge for the group. The agents in the group have to pool their knowledge together in order to obtain what is implicitly known.

A question, which is posed among agents, can be (partially) answered only if it is at least implicitly (partially) answered by a group. The next section shows one of the ways of communication formalization in the role of such 'answer mining'.

4.3 PUBLIC ANNOUNCEMENT

Let us show the role of public announcement using the familiar group of three agents—Anne, Bill, and Catherine. It is group knowledge that each of them has one card and nobody knows the cards of the others and that one of the cards is the Joker. Anne received the Joker, but neither Bill nor Catherine know which of the other two friends has it. In particular, both of them are not able to distinguish between the states where Anne has the Joker and where she has not. If Anne publicly announces

I've got the Joker,

everybody in the group learns this fact. Possible worlds (states) where Anne does not have the Joker are excluded from the (epistemic) models of both Bill and Catherine. Our example gives a typical situation represented in the public announcement logic—after a public announcement of a statement φ (here *I've got the Joker*), some other statement ψ holds, e.g., *Bill knows Anne has the Joker and Catherine knows Anne has the Joker.*

In fact, the author of an announced statement is irrelevant in our framework. The statement is understood as information coming to each member of a group in the same way. From this viewpoint, Anne's announcement has the same effect as if an external observer publicly announces *Anne has the Joker* to the group. Nevertheless, we can understand both announcements a different way. In the case that a member of a group announces a sentence, we can consider this announcement yielding more information than only an announced 'fact', it can explicitly publish the agent's knowledge of the 'fact'.

Formally, we introduce the logic of public announcement as an extension of system S5, cf. [48]. We define a box-like operator [ ], such that the intended meaning of the formula $[\varphi]\psi$ is:

After the public announcement of φ, it holds that ψ.

The semantics of the new operator is given in the following clause:

- $(\mathbf{M}, s) \models [\varphi]\psi$ iff $(\mathbf{M}, s) \models \varphi$ implies $(\mathbf{M}|_\varphi, s) \models \psi$

where $\mathbf{M}|_\varphi = \langle\langle S', R'_1, \dots, R'_m\rangle, v'\rangle$ and

$$\begin{aligned} S' &= \{s \in S \mid s \models \varphi\} \\ R'_i &= R_i \cap (S' \times S') \\ v'(p) &= v(p) \cap S' \end{aligned}$$

The model $\mathbf{M}|_\varphi$ is obtained from $\mathbf{M}$ by deleting all states where φ is not true and by the enforced restrictions of accessibility relations and the valuation function. Again, we can introduce a dual operator $\langle\ \rangle$ defined in a standard way as $\langle\varphi\rangle\psi$ iff $\neg[\varphi]\neg\psi$. If we rewrite the corresponding semantic clause, we obtain

- $(\mathbf{M}, s) \models \langle\varphi\rangle\psi$ iff $(\mathbf{M}, s) \models \varphi$ and $(\mathbf{M}|_\varphi, s) \models \psi$

The meaning of the dual operator is 'after a *truthful* announcement of φ, it holds that ψ'. It is easy to see that the diamond-like operator is stronger in the following sense:

Proposition 30. *$\langle\varphi\rangle\psi \to [\varphi]\psi$ is valid.*

Language $\mathcal{L}^{K[]}_{cpl}$ has the same expressive power as language $\mathcal{L}^{K}_{cpl}$. This is demonstrated by the next proposition, which provides a reduction of formulas with the public announcement operator to the epistemic ones. The corresponding equivalences give, in fact, an axiomatization of the announcement operator in public announcement epistemic logic without common knowledge, cf. [48, p. 81].

Proposition 31. *The following equivalences are valid formulas in* S5 *with public announcement modality, where $\circ \in \{\wedge, \vee, \to\}$:*

$$\begin{aligned} [\varphi]p &\leftrightarrow (\varphi \to p) \\ [\varphi]\neg\psi &\leftrightarrow (\varphi \to \neg[\varphi]\psi) \\ [\varphi](\psi \circ \chi) &\leftrightarrow ([\varphi]\psi \circ [\varphi]\chi) \\ [\varphi][i]\psi &\leftrightarrow (\varphi \to [i][\varphi]\psi) \\ [\varphi][\psi]\chi &\leftrightarrow [\varphi \wedge [\varphi]\psi]\chi \end{aligned}$$

For the common knowledge operator there is no such reduction (axiom), language $\mathcal{L}_{cpl}^{KC[]}$ is more expressive than $\mathcal{L}_{cpl}^{KC}$ [48, p. 232]. We only have a rule describing how a formula becomes common knowledge after a public announcement [48, p. 83]:

$$\frac{(\chi \wedge \varphi) \rightarrow [\varphi]\psi \wedge E_G\chi}{(\chi \wedge \varphi) \rightarrow [\varphi]C_G\psi} \tag{4.2}$$

The rule will be useful in the solution of an important question in public announcement logic:

What formulas become common knowledge after being announced?

We will deal with it in the next section.

For now, our formal work will proceed in the rich modal language $\mathcal{L}_{cpl}^{KECDQ[]}$ with formulas defined in BNF as follows:

$$\begin{aligned} \varphi \quad ::= \quad & p \mid \neg\psi \mid \psi_1 \vee \psi_2 \mid \psi_1 \wedge \psi_2 \mid \psi_1 \rightarrow \psi_2 \mid \psi_1 \leftrightarrow \psi_2 \mid \\ & [i]\psi \mid E_G\psi \mid C_G\psi \mid D_G\psi \mid \\ & ?_i\{\psi_1, \dots, \psi_n\} \mid ?_G\{\psi_1, \dots, \psi_n\} \mid \\ & [\psi_1]\psi_2 \end{aligned}$$

We have obtained public announcement logic with common knowledge, distributed knowledge, and questions: PACDQ.[5]

4.3.1 UPDATES AND QUESTIONS

As we said, members of a group learn what was announced. In particular, if Anne says

I've got the Joker,

the announced fact becomes *commonly known* in the group of players (Anne, Bill, Catherine). This seems to suggest that a publicly announced proposition becomes common knowledge . But what if Anne says:

You don't know it yet, but I've got the Joker.?

This announcement can be formalized by

$$J_a \wedge \neg[b](J_a) \wedge \neg[c](J_a),$$

where J_a means *Anne has the Joker*. Although the formula is true at the moment of announcement, it is evident that its epistemic part (*you don't know it yet*)

5 From now on, we will work in this system if not stated otherwise.

becomes false after it is announced. So the formula $(J_a \wedge \neg[b](J_a) \wedge \neg[c](J_a))$ becomes false after the announcement, i.e., in the updated model.

A formula which becomes false after it is truthfully announced (as in our example) is called an *unsuccessful update*; if it becomes true, we call it a *successful update*.

Definition 31.

- *Formula φ is a* successful update *in* $(\mathbf{M}, s)$ *iff* $(\mathbf{M}, s) \vDash \langle\varphi\rangle\varphi$.
- *Formula φ is an* unsuccessful update *in* $(\mathbf{M}, s)$ *iff* $(\mathbf{M}, s) \vDash \langle\varphi\rangle\neg\varphi$.

If a formula is an unsuccessful update, it cannot be commonly known in the updated model. We can prove that a formula is true after a public announcement if and only if it becomes common knowledge after the announcement; use the soundness proof of rule (4.2).

Proposition 32. $[\varphi]\psi$ *is valid iff* $[\varphi]C_G\psi$ *is valid.*

As a consequence we obtain[6]

Proposition 33. $[\varphi]\varphi$ *is valid iff* $[\varphi]C_G\varphi$ *is valid.*

If formula $[\varphi]\varphi$ is valid, we call it a *successful formula*.

Definition 32. *Formula φ is a* successful formula *iff* $[\varphi]\varphi$ *is valid, otherwise it is an* unsuccessful formula

From Proposition 33 we know that publicly announced successful formulas are commonly known. Successful formulas are preserved under a 'deleting states' process, i.e., they are preserved in submodels. Atoms are most obvious examples of successful formulas.[7]

Successful formulas which are true in a state are successful updates there [48, p. 85]:

Proposition 34. *If* $[\varphi]\varphi$ *is valid formula and* $(\mathbf{M}, s) \vDash \varphi$, *then* $(\mathbf{M}, s) \vDash \langle\varphi\rangle\varphi$.

It is easy to verify that questions are successful formulas in S5-systems.

Proposition 35. $[Q^i]Q^i$ *is valid.*

Proof. It is sufficient to prove that no 'cutting of' states in the pointed model $(\mathbf{M}, s)$ forced by the truthful public announcement of Q^i results in $(\mathbf{M}|_{Q^i}, s) \nvDash Q^i$. In S5-models, a question Q^i, which is askable in a state s, is askable in all states from the equivalence class sR_i. If $Q^i = ?_i\{\alpha_1, \dots, \alpha_n\}$ is askable in s, then three conditions are satisfied:[8]

6 For proofs of Proposition 33 and the previous one, see [48, p. 83 and 86].

7 Other examples are discussed in [48, pp. 86–88].

8 In the third condition, we will use directly the prospective presupposition of Q^i.

1. $s \models \neg[i]\alpha$, for each $\alpha \in dQ^i$
2. $s \models \langle i \rangle\alpha$, for each $\alpha \in dQ^i$
3. $s \models [i] \bigvee_{\alpha \in dQ^i} \alpha$

Whenever we suppose, for contradiction, that there is a state $t \in sR_i$ where $t \not\models Q^i$, it means that at least one of the three conditions is not true in state t. It is, $t \models [i]\alpha$, for some $\alpha \in dQ^i$, or $t \models [i]\neg\alpha$, for some $\alpha \in dQ^i$, or $t \models \langle i \rangle \bigwedge_{\alpha \in dQ^i} \neg\alpha$. However, it is not possible, because state t is in the after set sR_i and $tR_i = sR_i$. □

As a corollary of the proposition, we obtain that formula $Q^i \leftrightarrow [i]Q^i$ is valid. What is important for a public announcement communication is that a publicly announced question is commonly known among agents (see Proposition 33).

Successful formulas have an important property: they do not bring anything new if they are announced repeatedly.

Proposition 36. *Let φ be a successful formula. $[\varphi][\varphi]\psi \leftrightarrow [\varphi]\psi$ is valid.*

Proof. $[\varphi][\varphi]\psi$ is equivalent to $[\varphi \wedge [\varphi]\varphi]\psi$ (Proposition 31), which is equivalent to $[\varphi]\psi$, because of the validity of $[\varphi]\varphi$ (because φ is successful). □

It is no surprise that askable questions (in a state) are successful updates; this follows from Proposition 34 and Proposition 35:

Fact 38. $(\mathbf{M}, s) \models Q^i$ *iff* $(\mathbf{M}, s) \models \langle Q^i \rangle Q^i$.

4.3.2 PUBLIC ANNOUNCEMENT AND ANSWERHOOD

In the example we use throughout the chapter, Anne has the Joker, but neither Bill nor Catherine know it. If Catherine publicly asks[9]

> *Who has got the Joker?,*

Bill can infer:

> *I have not got the Joker and Catherine does not know who has it, therefore Anne has it.*

Catherine's question was *informative* for Bill in this situation. Now, the question *Who has got the Joker?*, which was askable for Bill, became non-askable after Catherine had asked it. Even though her question was not (partially) publicly answered. The presented idea is behind the definition of *informative formula*.

Definition 33 (Informative formula). *A formula φ is* informative *for agent i with respect to question Q in* $(\mathbf{M}, s)$ *iff* $(\mathbf{M}, s) \models Q^i \wedge \langle\varphi\rangle\neg Q^i$.

9 Catherine's publicly asked question announces that the question is askable for her.

Contrary to partial answerhood (see Proposition 26) there need not be any logical connection between an informative formula and direct answers to a question. The informativeness can be forced by the form of a particular model.

However, it is a clear conclusion of Definition 33 that whenever there is an askable question for an agent in a state, then after an announcement of an informative formula the agent obtains at least a partial answer to the question.

Proposition 37. *If a formula φ is informative in $(\mathbf{M}, s)$ for an agent i with respect to a question Q, then there is $\alpha \in dQ$ such that $(\mathbf{M}|_{\varphi}, s) \models [i]\alpha$ or $(\mathbf{M}|_{\varphi}, s) \models [i]\neg\alpha$.*

Proof. From the informativeness of φ we obtain $(\mathbf{M}|_{\varphi}, s) \models \neg Q^i$, which means either $(\mathbf{M}|_{\varphi}, s) \models A_i Q$ or $(\mathbf{M}|_{\varphi}, s) \models P_i Q$. The askability of Q^i in $(\mathbf{M}, s)$ (Definition 33) causes the question to be safe in $(\mathbf{M}, s)$ and cannot be weakly presupposed. □

We have to be careful—informativeness of a formula for one agent does not imply its informativeness for another one. The next example shows a structure where φ is informative for the agent a with respect to the question $?\alpha$, but it is not informative for b with respect to the same question:

Example 21.

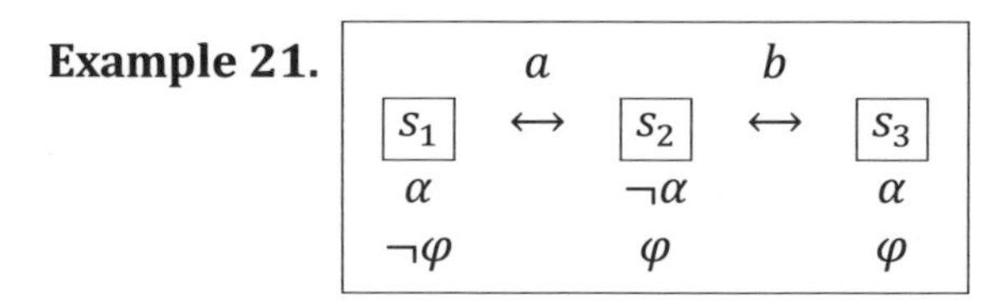

4.3.3 ANSWER MINING IN A GROUP

In Example 20, we considered two agents, a and b. Neither of them can answer the question $?(\alpha \to \beta)$, it is their group question. However, the question is implicitly answered. Let us imagine their possible cooperative communication. The question $?(\alpha \to \beta)$ is askable for the agent a and she wants to find an answer to it. Her colleague b might help. The agent a can directly and publicly ask the question $?(\alpha \to \beta)$ and reveal her ignorance in the group $\{a, b\}$. A question is a successful formula, thus, it is commonly known to the group. In cooperative communication, agents are supposed to announce what they know with respect to their group question, i.e., a can provide the information that α is known by her, $[a]\alpha$, and b can express his knowledge about $\neg\beta$, $[b]\neg\beta$. Finally, they reach the (complete) answer $\neg(\alpha \to \beta)$, which will be, moreover, common knowledge among them.

We obtained a sequence of (truthfully) publicly announced agents' knowledge leading to a (commonly known) answer:

$$\langle [a]\alpha \rangle \langle [b]\neg\beta \rangle \neg(\alpha \to \beta),$$

respectively,

$$[[a]\alpha][[b]\neg\beta] C_{\{a,b\}} \neg(\alpha \to \beta).$$

The group of agents $\{a, b\}$ was successful in seeking an answer using just 'internal' resources. They reached an answer after a series of announcements of facts they knew.

We would like to generalize this example. The idea behind could be the following:

> There is a group question, but an answer is hidden in a form of distributive knowledge. The answer should be reachable by (finite) communication based on public announcements in the group. These public announcements provide successful updates that terminate in the commonly known answer to the initial group question.

Nonetheless, the idea presented brings many problems here. First, the definition of distributive knowledge does not provide any finite way to obtain such hidden knowledge.[10] Second, in the example the agents communicated, in fact, their knowledge of atomic formulas, i.e., they announced successful formulas and updates. An ideal form of public announcement communication is that announcements are commonly known (successful formulas) and true in updated states (successful updates).

The solution to the first problem is inspired by rule (4.1) where agents $l_1, \ldots, l_k$ pool their knowledge together to obtain implicit knowledge. Simultaneously, we were inspired by an alternative definition of distributive knowledge presented in paper [13].[11]

Definition 34 (Finite distributive knowledge). *$(\mathbf{M}, s) \vDash D^f_G \varphi$ iff there are formulas $\psi_1, \ldots, \psi_k$ such that $(\mathbf{M}, s) \vDash [l_1]\psi_1 \wedge \ldots \wedge [l_k]\psi_k$, where $l_1, \ldots, l_k$ are agents from G and formula $(\psi_1 \wedge \ldots \wedge \psi_k) \to \varphi$ is valid.*

Fact 39. *If $(\mathbf{M}, s) \vDash D^f_G \varphi$, then $(\mathbf{M}, s) \vDash D_G \varphi$.*

In the proof of this fact, we can use the same argument as in [13]. Let us suppose that $(\mathbf{M}, s) \vDash D^f_G \varphi$, then there is a state $t \in \left(s \bigcap_{i \in G} R_i\right)$, or not. If not,

10 Moreover, we work in a non-compact system due to the common knowledge operator, cf. [48, p. 182].

11 $(\mathbf{M}, s) \vDash D^*_G \varphi$ iff $\{\psi \mid (\exists i \in G)((\mathbf{M}, s) \vDash [i]\psi)\} \vDash \varphi$. D^*_G is not equivalent to D_G in all systems: If $(\mathbf{M}, s) \vDash D^*_G \varphi$, then $(\mathbf{M}, s) \vDash D_G \varphi$, but not vice versa [13]. Our definition of *finite distributive knowledge* is not equivalent to D^*_G thanks to non-compactness.

then $(\mathbf{M}, s) \models D_G\varphi$ for any formula φ. If yes, then formula $(\psi_1 \wedge ... \wedge \psi_k)$ is true in t, so is φ.

On the notion of finite distributive knowledge, we base the idea of finite *solvability*.

Definition 35 (Solvability). *A question Q is*

- solvable *for a group G in a state $(\mathbf{M}, s)$ iff there is a formula φ such that $(\mathbf{M}, s) \models D_G^f\varphi$, and $[l_1]\psi_1 \wedge ... \wedge [l_k]\psi_k$ is a successful formula, and $\varphi \to \alpha$ is valid for some $\alpha \in dQ$,*
- partially solvable *for a group G in a state $(\mathbf{M}, s)$ iff there is a formula φ such that $(\mathbf{M}, s) \models D_G^f\varphi$, and $[l_1]\psi_1 \wedge ... \wedge [l_k]\psi_k$ is a successful formula, and $\varphi \to \neg\alpha$ is valid for some $\alpha \in dQ$.*

The definition says that hidden information can provide at a least partial answer to a question in finite steps done by pooling knowledge together. We want to do it by public announcement communication, so, the important clause is that the formula pooling agents' knowledge together is successful. This condition ensures that the distributive knowledge becomes common knowledge in the updated model and, simultaneously, that the series of updates based on the individual knowledge of agents terminates in the same result.

Now we show that a (partially) solvable question can be (partially) answered in a group of agents, i.e., a (partial) answer becomes common knowledge in the group.

Proposition 38. *If a question Q is (partially) solvable for a group of agents G in a state $(\mathbf{M}, s)$, then there is a direct answer $\alpha \in dQ$ such that α (resp. $\neg\alpha$) is commonly known after $\bigwedge_{j=1}^k [l_j]\psi_j$ is announced.*

Proof. It is sufficient to prove: if $(\mathbf{M}, s) \models D_G^f\varphi$ and $\bigwedge_{j=1}^k [l_j]\psi_j$ is successful, then $(\mathbf{M}|_{\bigwedge_{j=1}^k [l_j]\psi_j}, s) \models C_G\varphi$, where φ plays the role as in Definition 35.

First, let us prove $[\bigwedge_{j=1}^k [l_j]\psi_j]C_G\varphi$. Since $\bigwedge_{j=1}^k [l_j]\psi_j$ is successful, formula $[\bigwedge_{j=1}^k [l_j]\psi_j]C_G(\bigwedge_{j=1}^k [l_j]\psi_j)$ is valid. Together with the validity of the formula $(\psi_1 \wedge ... \wedge \psi_k) \to \varphi$ we obtain the validity of $[\bigwedge_{j=1}^k [l_j]\psi_j]C_G\varphi$.

Second, we have to prove that being $\bigwedge_{j=1}^k [l_j]\psi_j$ truthfully announced in a state $(\mathbf{M}, s)$, the formula φ is commonly known in the updated state $(\mathbf{M}|_{\bigwedge_{j=1}^k [l_j]\psi_j}, s)$. This follows from two facts:

- φ is distributive knowledge in $(\mathbf{M}, s)$,
- $\bigwedge_{j=1}^k [l_j]\psi_j$ is true in $(\mathbf{M}, s)$.

□

4.4 PUBLIC ANNOUNCEMENT LOGIC WITH QUESTIONS (FINAL REMARKS)

This chapter combines questions introduced in an epistemic framework (Chapter 3) and communication based on public announcement logic. The background system was multi-modal S5, which is 'introspective' (axioms 4 and 5) and 'factive' (axiom T) in its epistemic interpretation.[12] The presented extension of this system, *public announcement logic* with group knowledge operators, serves as a model of specific communication. The communication is modeled by cutting off states (and accessibility relations) from an initial epistemic model. This leads to a new epistemic model, which is a submodel of the previous one. We introduced the idea that communication among agents can be understood as a series of public announcements. The central task for public announcement logic is when an announced formula becomes common knowledge in a group of agents, see, e.g., [48]. This role is performed by successful formulas. In accordance with it, we believe it is necessary to share communication results in a group, this means, agents share announced questions as well as knowledge exchanged among them. We saw that questions were successful formulas, but in the case of the series of announced individual knowledge we had to be careful. On the one hand, if there is an exchange of known facts, it can be considered as a series of successful updates. On the other hand, generally, a series of successful formulas need not terminate in the required form—that was the reason we required in Definition 35 that the whole conjunction of agents' knowledge was a successful formula.[13] Although it seems rather awkward to formalize the process of communication by public announcement logic, which is a good formalization for 'one-step updates', we found some interesting and perhaps useful concepts there. Knowledge modalities for groups of agents provide new answerhood conditions. Common knowledge is important for sharing information in a group and forms the answerhood condition for a question *being (partially) answered among agents.* (Finite) distributed knowledge can formalize hidden information, which is under some conditions reachable, see Propositions 29 and 38.

12 S5 system includes all theorems of CPL (with modus ponens rule) and modal axiomatic schemas for each agent $i \in \mathcal{A}$: $[i](\varphi \rightarrow \psi) \rightarrow ([i]\varphi \rightarrow [i]\psi)$ (axiom K), $[i]\varphi \rightarrow \varphi$ (axiom T), $[i]\varphi \rightarrow [i][i]\varphi$ (axiom 4), $\neg[i]\varphi \rightarrow [i]\neg[i]\varphi$ (axiom 5), and the necessitation rule $\frac{\varphi}{[i]\varphi}$.

13 As a consequence we have that each conjunct of $([l_1]\psi_1 \wedge \ldots \wedge [l_k]\psi_k)$ is a successful formula.

5. CONCLUSION

5.1 THE STARTING POINT: SAM

Though the book consists of two almost independent parts, some points are common to both of them. Working in *inferential erotetic logic* (Chapter 2) as well as in *erotetic epistemic logic* (Chapters 3 and 4) we primarily wanted to provide tools for the development of both branches of erotetic logic. Even if we do not want to contribute to a philosophical debate on what a question is, it seems to us that a word or two should be said about the chosen set-of-answers methodology applied in both parts of the text.

The main inspiration came from the original inferential erotetic logic and, simultaneously, from intensional erotetic logics. They inspired the set-of-answers methodology introduced in the first chapter, as well as the emphasis posed on inferences with questions.

The idea of SAM is liberal enough to be used in the intended applications of presented approaches. It is open to additional syntactical as well as semantic restrictions. Of course, many objections can be raised against it, especially, whenever we want to analyze all kinds of natural language questions. Both IEL and erotetic epistemic logic want to profit maximally from the background 'logic of declaratives'. Above that, SAM proved its convenience for the formalization of erotetic inferences. In the form presented in Chapter 2, SAM makes it possible to 'erotetize' consequence relations. This setting mixes questions and declaratives only on the level of inferences. In IEL we tried to keep the first two of Hamblin's postulates (cf. Section 1.2.1). On the contrary, erotetic epistemic logic (Chapters

3 and 4) admits to having not only declaratives among direct answers. This was important for 'questions about questions' and for the analysis of *I don't know* answers. We did not study this problem in the chapters on epistemic logic, but we mentioned it in the last paragraph of Section 1.4.

5.2 MAIN RESULTS

The first part (Chapter 2) is fully developed in the IEL framework with the significance of SAM formalization of questions. The chosen SAM and model-based approach make it possible to model questions by their *semantic range*.[1] The benefit is that some properties and relationships appear more transparent. Our interest and study was primarily focused on inferences with questions. Many theorems proved in Chapter 2 provide similar or the same results as the theorems of the original inferential erotetic logic. However, we often obtained the results in another way.[2] Let us now list some interesting and new results that can be understood as our contribution to the development of IEL variants:

- The semantic range of questions plays a role in *pure erotetic implication*. The relation of pure e-implication between two questions causes both questions to have the same semantic range (Proposition 6). As a consequence we obtain that classes of questions (normal and regular) are closed under pure e-implication (Propositions 8 and 9).
- We were exceptionally interested in relationships among consequence relations with questions. Proposition 12 shows that *evocation* is preserved under *regular e-implication*.
- The concept of *giving an answer* establishes similar relations. For example, Proposition 15 connects regular e-implication and evocation.
- In the last part of Chapter 2, we bring into play the *reducibility* of an initial question to a set of questions. We wanted to be general enough and combine erotetic consequences with the concept of reducibility. Regular e-implication and evocation (with respect to 'giving an answer') are connected with reducibility, see Propositions 19 and 20. In the last subsection of Chapter 2 (Reducibility and sets of yes-no questions), we studied the generalized conditions of the reducibility of an initial question to a set of yes-no questions. The established results correspond to the results in the original IEL, see the references in the subsection.

The second part (Chapter 3 and Chapter 4) is different in its substance. The epistemic framework forms the background, and a question corresponds to a

1 The inspiration comes from a mixture of IEL and partitioning concept in Groenendijk-Stokhof's approach, cf. Section 1.3.2.
2 References can be found in the Chapter 2.

certain state of ignorance and presuppositions of an agent. This approach is a novelty inspired by epistemic terminology often used by erotetic logicians.[3] Questions become a part of epistemic language and they can be considered as 'satisfied' in an epistemic state (of a model). Questions' truth conditions in a state of a Kripke model (we use the term *askability* in the state) are more complex than in the case of declaratives, and they are based on three items (*non-triviality*, *admissibility*, and *context*). Although we introduced a general epistemic-like approach in Chapter 3, we have decided then to use a finite variant of SAM. The decision corresponds to our idea of 'well-arranged' communications. At the beginning, we had also wanted to provide a very general approach to questions' formalization in epistemic framework, but we recognized that we had to be cautious of it. In the end, as an appropriate background formalization for public announcement logic with questions the multi-modal logic S5 was chosen, which is common in this area (Chapter 4). Some logicians might consider S5 undesirable. Irrespective of objections against epistemic properties (e.g., the logical omniscience problem), our questions' analysis is 'reductionism' in its nature—questions are representable by a complex modal formulas in S5. However, it has some advantages. We can fully exploit the properties of public announcement logic, in particular, successful formulas and updates that can be problematic and sometimes must be redefined in weaker systems.

Chapter 3 tries to develop the basics of *erotetic epistemic logic*:

- The starting erotetic term in the single-agent epistemic framework is the *askability* of a question by an agent in a state. This concept establishes individual answerhood conditions, i.e., conditions under which questions are completely or partially answered for an agent in a given state. The relationship of both partial and complete answerability depends on the form of a set of direct answers (Fact 30 and Proposition 26). As a result we have questions, which meet Hamblin's third postulate, that are prominent.
- In building the epistemic logic of questions, we tried to find similar erotetic notions as we have in the IEL framework. We defined the *presuppositions* and *prospective presuppositions* of a question (Section 3.2.1). The term *safe question* in global as well as local form is based on it (Definitions 21 and 22).
- Inferential structures with questions are defined with the help of the standard implication (Section 3.3). In Section 3.5 we studied the properties of 'erotetic implications' with respect to a set of auxiliary formulas (relativized askability) and, in the last section (3.6), we introduced a variant of 'reducibility' of a question to a conjunction of yes-no questions.

Multi-agent epistemic logic with questions in Chapter 4 introduces group erotetic terms.

3 The semantic approach is inspired by Groenendijk's and Stokhof's papers.

- We defined the askability of a question for a group of agents in a state (Definition 29) and used group knowledge operators to establish group answerhood conditions. *Implicit answerability*, stated in Definition 30, was based on the distributive knowledge operator. Common knowledge is then important in the case of the answerability of a question for a group of agents.
- The dynamic step takes place in the framework of public announcement logic (built over the modal logic S5). We proved that
  - *publicly asked questions* are successful formulas, so they are commonly known after being publicly announced (Proposition 35),
  - an askable question in a state is a successful update there (Fact 38).
- In addition to that we introduced a concept of formula informative for an agent with respect to a question (Definition 33), which is a form of answerability after an update.

5.3 WEAK POINTS AND PROBLEMS

5.3.1 SAM AND CONTEXT CONDITION OF ASKABILITY

One of the problematic points is the introduced *set-of-answers methodology*. Critics have aimed at the impossibility of such a representation to model all kinds of questions. Our effort was primarily to build transparent communication among agents. This led us to a version of SAM where questions have finite sets of direct answers. However, there are many questions that have well defined infinite sets of direct answers. For example, a question

Which natural number is greater than 13?

can require an answer from an infinite but well defined set of answers (natural number(s) greater than 13). Our intention to provide complex information on the ignorance and expectations of an agent, publicly asking a question, connect the problem of SAM with the concept of *askability* in *erotetic epistemic logic*. The expectation of an agent is mirrored in the third condition of Definition 20 (context condition) which requires that, in a finite SAM, an agent (explicitly) knows the disjunction of all direct answers. This places a restriction on the epistemic possibilities of an agent-questioner. Nevertheless, agents often ask questions without being aware of all possible direct answers.[4] We met this problem in an application of erotetic epistemic logic in [49] and resigned ourselves to the *context* condition. However, we feel the need to discuss the role of 'context' as

4 Personal communication with Andrzej Wiśniewski and Igor Sedlár. Igor mentioned a question *Where is Bill?* Asking it, an agent need not know all the possibilities of Bill's position.

shared knowledge in a group of agents. We believe that this (commonly known) 'context' forms an explicit knowledge framework for agents in a group. It means that agents communicate just in the borders delimited by 'context'. In cooperative communication they are obliged not to leave these borders.

5.3.2 ASKABILITY AND BELIEF

Our original ambition of the full generality of the *askability* concept was exaggerated. We wanted to use it in any epistemic-like logic, but there are examples of questions that are not treatable in our framework, namely, questions aiming at a revision of one's beliefs:[5]

> Suppose an agent i has the following beliefs $[i]\varphi_1, \dots, [i]\varphi_n$ in a state. If someone says *One of your beliefs is false*, it is natural for i to ask *Which one is it?*, but this question $?_i\{\neg\varphi_1, \dots, \neg\varphi_n\}$ is not askable in the state.

That is the reason we have to be careful to use the definition of askability in other formal systems.

Answerhood conditions based on *askability* can bring non-compliance with one's intuition in the remaining conditions as well. The concept of partial answerhood dependent on the *admissibility* condition (Definition 20) is an example. A partially answered question is not 'solved' and a questioner could want to ask it until she receives a complete answer. But we say that such question is non-askable.[6] In our conception of dynamic epistemic approach, a questioner that receives a partial answer to a question is obliged to pose a slightly changed question, which is now askable and reflects the received partial answer.

5.3.3 INFERENCES WITH QUESTIONS

We used inferences with questions as an appropriate justification of the development of erotetic logic. The inferential structures we demonstrated were closely related to consequence relations in *inferential erotetic logic*, which we understand as an important inspiration for our approach. IEL-like consequences form a contact point between Chapter 2 and Chapter 3. Above that, the epistemic framework of Chapter 3 provides another kind of inference with questions:[7]

---

5 Igor Sedlár put forward this example.
6 Personal communication with Andrzej Wiśniewski.
7 Personal communication with Igor Sedlár.

> Whenever an agent a asks an agent b a question, the agent b can *infer* that a does not have (know) any answer to the question.

This is, in fact, the original idea of communication based on the posing of questions. Our concept of askability suggests what kind of information is delivered to an agent-listener. A listener is aware of a questioner's ignorance as well as epistemic possibilities.

5.3.4 QUESTIONS AMONG DIRECT ANSWERS

As a corollary of Proposition 35, we obtained that a question Q is askable in a state by an agent i if and only if i knows Q, i.e., in S5

$$(\mathbf{M}, s) \models Q^i \text{ is equivalent to } (\mathbf{M}, s) \models [i]Q^i$$

Thus, if Q is askable by i, then the question

> Is Q askable?

is *not askable* by i.[8] The question *Is Q askable (by i)?* can be formalized as a yes-no question with Q in the role of direct answers

$$?_i\{Q^i, \neg Q^i\}.$$

We discussed similar structures in Section 1.4.1. In systems weaker than S5 it is possible to have a non-askable question Q (by i) in a state and, simultaneously, the question *Is Q askable (by i)?* is askable (by i) there.

We do not mention this kind of question as a weak point of our approach. We believe that it is a good trait.

5.4 RELATED WORK

Publications related to *inferential erotetic logic* were mentioned mainly in Chapter 2. We list some papers that have to do something with an 'inquiry' aspect of IEL in Section 1.3.1 because this was not studied in Chapter 2. In Section 1.3.1 we also added two papers based on Hintikka's approach, which can be understood as a rival to IEL. The complete survey of the original IEL, with an emphasis on generality, can be found in the nice book [59] by Andrzej Wiśniewski. If someone is interested in the philosophy of erotetic logic, we recommend some essays in [58].

---

8 Igor Sedlár drew attention to it.

Section 1.3.2 includes many publications referring to the *intensional erotetic logic* of Groenendijk and Stokhof. Their approach was an inspiration for the dynamization of epistemic logic with questions. A notable work was done by Ştefan Minică and Johan van Benthem [46]. This paper makes a great step toward a combination of dynamic logic with questions and public announcement. The usual epistemic model is enriched by a new equivalence relation of indistinguishability (*abstract issue relation*). Similarly to updates of models based on public announcements, there are updates for asking yes-no questions in the propositional version. Roughly speaking, a publicly announced question $?\alpha$ produces partitioning of the logical space into two parts, one with α-states and the other one with $\neg\alpha$-states, it means that two issues are distinguished.[9] The behaviour of publicly announced questions is almost the same as the behaviour of publicly announced declaratives. Question-updates delete the abstract issue relation, and declarative-updates delete the epistemic indistinguishability relation.

Another inspiration for the dynamic logic of questions is *inquisitive semantics* developed by Jeroen Groenendijk (see [15]) and his collaborators Ivano Ciardelli and Floris Roelofsen. The generalized version with its associated logic can be found in [5]. From the dynamic point of view, propositions are seen as proposals on how to update the common state.

> "If a proposition consists of two or more possibilities, it is *inquisitive*: it invites the other participants to provide information such that at least one of the proposed updates may be established." [5, p. 112]

In inquisitive logic, the formula $?\alpha$ is defined as $(\alpha \vee \neg\alpha)$ and it is an example of a question; in fact, it invites us to provide information on whether α or $\neg\alpha$ (the formula $?\alpha$ is *inquisitive*) and, simultaneously, it does not exclude some of the states (the formula is not *informative*).[10] The connection between inquisitive semantics and inferential erotetic logic is presented in paper [60], which can be read as an introduction to inquisitive semantics as well.

In the first chapter we mentioned that there are some reductionist theories. One of them is *transparent intensional logic* (TIL) developed by Pavel Tichý. On the one hand, questions have the same semantic core as their adequate answers on the semantic level and the act of questioning should be analyzed on the level of pragmatics [43, 34, 9]. On the other hand, in recent publications, TIL provides a tool for analyzing some erotetic concepts used, especially in communication among agents, see, e.g., [10]. Communication analysis seems to make the bridge between different approaches.

---

9 The abstract issue relation is deleted between states where α is true and states where $\neg\alpha$ is true.

10 Only non-informative formulas are called *questions*.

David Harrah's papers, which we have cited in the book, could be considered a 'metatheory' of erotetic logic. In many cases, they describe a history of questions' formalizations in logic, and the general properties that are independent of the fine differences among the logic of questions. Recently, Polish erotetic logicians have published some contributions to erotetic 'metatheory'. Let us mention two of them that can be considered as having consequences for artificial intelligence. Paper [61] contains a strengthening of Harrah's incompleteness theorem by tools of recursion theory.[11] The other paper [25] presents the *Turing interrogative game* (TIG) as an instance of the Turing test. TIG is an exchange of questions and answers between 'interrogator' and 'machine'. If some conditions hold, a 'machine' (no matter whether thinking or not) loses a certain TIG.

5.5 FUTURE DIRECTIONS (SOME OF THEM)

As recent publications indicate, the combination of epistemic and dynamic aspects seems to be a good framework for the development of erotetic logic. The posing of questions, as well as answering, is a natural form of update in communication. In this book we presented a way to incorporate questions into the epistemic framework and demonstrated just a small step toward dynamization. We hoped to provide very general approach. However, we were not completely successful (see Section 5.3.2). We are obliged to revise the term *askability*, together with the used set-of-answers methodology. *Context condition* is probably the weak point of the definition of *askability*. Nonetheless, in Section 5.3.1 we mentioned that some concept of 'context' seems important to us in epistemic logic with group modalities. The revision of finite SAM is linked with a deeper study of a broader set of questions (and their answers) that can be analyzed, for example, in the framework of predicate logic.

We have just admitted that our 'dynamic step' was small. Thus, we would like to incorporate questions into richer dynamic systems. Action model logic with questions has been prepared, but not published yet. Other systems of logic of communication are very promising, see [47] as a representative example. Simultaneously, we have been thinking of other epistemic logic. Relevant (or, generally, substructural) epistemic logic can be taken into account, cf. [2, 26]. Relevant implication seems to provide a good background for erotetic implication. However, in the first instance, we have to develop a multi-agent version of substructural epistemic logic.

---

11 The core of the theorem is whether any set of sentences is the set of direct answers to some question. Beyond that, there are some epistemic consequences in the paper.

Another inspiration is non-monotonic logic. Non-monotonic approaches direct our attention to inference relations. In this context, IEL invites non-monotonic applications. Preferences among direct answers or default rules based on questions can be incorporated.

BIBLIOGRAPHY

[1] N. Belnap and T. Steel. *The Logic of Questions and Answers.* Yale, 1976.

[2] M. Bílková, O. Majer, M. Peliš, and G. Restall. Relevant agents. In L. Beklemishev, V. Goranko, and V. Shehtman, editors, *Advances in Modal Logic,* pages 22–38. College Publications, 2010.

[3] M. Borkent and R. van Rozen. On logic and questions in dialogue systems. http://www.pictureofthemoon.net/107borkent/olaqid.pdf (downloaded by June 2007).

[4] L. Běhounek. Fuzzification of Groenendijk-Stokhof propositional erotetic logic. *Logique & Analyse,* 185–188:167–188, 2004.

[5] I. Ciardelli and F. Roelofsen. Inquisitive Logic. In D. Grossi, L. Kurzen,and F. Velázquez--Quesada, editors, *Logic and Interactive Rationality,* pages 111–146. ILLC, 2010.

[6] E. Coppock. Toward a dynamic logic for questions, 2005. http://www.stanford.edu/~coppock/dynamic-questions.pdf (downloaded by April 2007).

[7] K. De Clercq and L. Verhoeven. Sieving out relevant and efficient questions. *Logique & Analyse,* 185–188:189–216, 2004.

[8] H. N. Duc. *Resource-Bounded Reasoning about Knowledge.* PhD thesis, Faculty of Mathematics and Informatics, University of Leipzig, 2001.

[9] M. Duží, B. Jespersen, and P. Materna. *Procedural Semantics for Hyperintensional Logic. Foundations and Applications of Trasnsparent Intensional Logic.* Springer, 2010.

[10] M. Duží, M. Číhalová, and M. Menšík. Communication in a multi-agentsystem; questions and answers. In *13th International Multidisciplinary Scientific GeoConference SGEM 2013,* pages 11–22, 2013.

[11] R. FAGIN, J. HALPERN, Y. MOSES, AND M. VARDI. *Reasoning about Knowledge*. MIT Press, 2003.

[12] E. J. GENOT. The game of inquiry: the interrogative approach to inquiry and belief revision theory. *Synthese*, 171:271–289, 2009.

[13] J. GERBRANDY. Distributed knowledge. In J. Hulstijn and A. Nijholt, editors, *Twendial'98: Formal Semantics and Pragmatics of Dialogue*, pages 111–124. Universiteit Twente, Enschede, 1998.

[14] J. GROENENDIJK. Questions and answers: Semantics and logic. In R. Bernardi and M. Moortgat, editors, *Proceedings of the 2nd CologNET-ElsET Symposium. Questions and Answers: Theoretical and Applied Perspectives*, pages 16–23. Universiteit Utrecht, 2003.

[15] J. GROENENDIJK. Inquisitive semantics: Two possibilities for disjunction. In P. Bosch, D. Gabelaia, and J. Lang, editors, *Logic, Language, and Computation: 7th International Tbilisi Symposium on Logic, Language, and Computation*, pages 80–94. Springer, 2009.

[16] J. GROENENDIJK AND M. STOKHOF. Partitioning logical space. Annotated handout, ILLC, Department of Philosophy, Universiteit van Amsterdam, 1990. Second European Summerschool on Logic, Language and Information; Leuven, August 1990; version 1.02.

[17] J. GROENENDIJK AND M. STOKHOF. Questions. In J. van Benthem and A. ter Meulen, editors, *Handbook of Logic and Language*, pages 1055–1125. Elsevier, Amsterdam, 1997.

[18] D. HARRAH. On the history of erotetic logic. In A. Wiśniewski and J. Zygmunt, editors, *Erotetic Logic, Deontic Logic, and Other Logical Matters*, volume 17, pages 19–27. Wydawnictwo Uniwersytetu Wrocławskiego, 1997.

[19] D. HARRAH. The logic of questions. In D. Gabbay and F. Guenthner, editors, *Handbook of Philosophical Logic*, volume 8, pages 1–60. Kluwer, 2002.

[20] J. HINTIKKA. *The Semantics of Questions and the Questions of Semantics*. North-Holland Publishing Company, 1976.

[21] J. HINTIKKA, I. HALONEN, AND A. MUTANEN. Interrogative logic as a general theory of reasoning. In D. Gabbay, R. Johnson, H. Ohlbach, and J. Woods, editors, *Handbook of the Logic of Argument and Inference*, pages 295–337. Elsevier, 2002.

[22] K. H. KNUTH. Toward question-asking machines: The logic of questions and the inquiry calculus, 2005. (Tenth International Workshop on Artificial Intelligence and Statistics (AISTATS 2005), Barbados) http://www.gatsby.ucl.ac.uk/aistats/fullpapers/246.pdf.

[23] P. LEŚNIEWSKI AND A. WIŚNIEWSKI. Reducibility of questions to sets of questions: Some feasibility results. *Logique & Analyse*, 173–175:93–111, 2001.

[24] D. LESZCZYŃSKA. Socratic proofs for some normal modal propositional logics. *Logique & Analyse*, 185–188:259–285, 2004.

[25] P. ŁUPKOWSKI AND A. WIŚNIEWSKI. Turing interrogative games. *Minds and Machines*, 21:435–448, 2011.

[26] O. MAJER AND M. PELIŠ. Epistemic logic with relevant agents. In M. Peliš, editor, *The Logica Yearbook 2008*, pages 123–135. College Publications, 2009.

[27] J. J. C. MEYER AND W. VAN DER HOEK. *Epistemic Logic for AI and Computer Science*. Cambridge University Press, 1995.

[28] R. NELKEN AND N. FRANCEZ. A bilattice-based algebraic semantics of questions. In *Proceedings of MOL6, Sixth Meeting on Mathematics of Language*, Orlando, Florida, 1999.

[29] R. NELKEN AND N. FRANCEZ. A calculus for interrogatives based on their algebraic semantics. In *Proceedings of TWLT16/AMILP2000: Algebraic methods in language processing*, pages 143–160, Iowa, 2000.

[30] R. NELKEN AND N. FRANCEZ. Querying temporal database using controlled natural language. In *Proceedings of COLING2000, The 18th International Conference on Computational Linguistics*, 2000.

[31] R. NELKEN AND N. FRANCEZ. Bilattices and the semantics of natural language questions, 2002.

[32] R. NELKEN AND C. SHAN. A logic of interrogation should be internalized in a modal logic for knowledge. In K. Watanabe and R. B. Young, editors, *SALT XIV: Semantics and Linguistic Theory*. Cornell University Press, 2004.

[33] R. NELKEN AND C. SHAN. A modal interpretation of the logic of interrogation. *Journal of Logic, Language, and Information*, 15:251–271, 2006.

[34] M. PELIŠ. Questions and logical analysis of natural language: the case of transparent intensional logic. *Logique & Analyse*, 185–188:217–226, 2004.

[35] M. PELIŠ. Consequence relations in inferential erotetic logic. In M. Bílková, editor, *Consequence, Inference, Structure*, pages 53–88. Faculty of Arts, Charles University, 2008.

[36] M. PELIŠ. *Logic of Questions*. PhD thesis, Faculty of Arts, Charles University in Prague, 2011.

[37] M. PELIŠ. Set of answers methodology in erotetic epistemic logic. *Acta Universitatis Carolinae – Philosophica et Historica*, 2/2010:61–74, 2013.

[38] M. PELIŠ AND O. MAJER. Logic of questions from the viewpoint of dynamic epistemic logic. In M. Peliš, editor, *The Logica Yearbook 2009*, pages 157–172. College Publications, 2010.

[39] M. PELIŠ AND O. MAJER. Logic of questions and public announcements. In *Eighth International Tbilisi Symposium on Language, Logic and Computation 2009*, pages 145–157. LNCS, Springer, 2011.

[40] C. SHAN AND B. TEN CATE. The partitioning semantics of questions, syntactically. In M. Nissim, editor, *Proceedings of the Seventh ESSLLI Student Session*, 2002.

[41] B. TEN CATE AND C. SHAN. Question answering: From partitions to prolog. In U. Egly and C. G. Fermüller, editors, *Proceedings of TABLEAUX 2002: Automated reasoning with analytic tableaux and related methods*, pages 251–265. Springer-Verlag, 2002.

[42] B. TEN CATE AND C. SHAN. Axiomatizing Groenendijk's logic of interrogation. In M. Aloni, A. Butler, and P. Dekker, editors, *Questions in Dynamic Semantics*, pages 63–82. Elsevier, 2007.

[43] P. TICHÝ. Questions, answers, and logic. *American Philosophical Quarterly*, 15:275–284, 1978.

[44] M. URBAŃSKI. Some remarks concerning modal propositional logic of questions. *Logic and Logical Philosophy*, 6:187–196, 1998.

[45] M. URBAŃSKI. Synthetic tableaux and erotetic search scenarios: Extension and extraction. *Logique & Analyse*, 173–175:69–91, 2001.

[46] J. VAN BENTHEM AND Ş. MINICĂ. Toward a Dynamic Logic of Questions. In X. He, J. Horty, and E. Pacuit, editors, *Logic, Rationality, and Interaction*, pages 27–41. Springer, 2009.

[47] J. VAN BENTHEM, J. VAN EIJCK, AND B. KOOI. Logics of communication and change. *Information and Computation*, 204:1620–1662, 2006.

[48] H. van Ditmarsch, W. van der Hoek, and B. Kooi. *Dynamic Epistemic Logic*. Springer, 2008.
[49] P. Švarný, O. Majer, and M. Peliš. Erotetic epistemic logic in private communication protocol. In M. Dančák and V. Punčochář, editors, *The Logica Yearbook 2013*, pages 223–238. College Publications, 2014.
[50] A. Wiśniewski. On the reducibility of questions. *Erkenntnis*, 40:265–284, 1994.
[51] A. Wiśniewski. *The Posing of Questions: Logical Foundations of Erotetic Inferences*. Kluwer, 1995.
[52] A. Wiśniewski. Erotetic logic and explanation by abnormic hypotheses. *Synthese*, 120:295–309, 1999.
[53] A. Wiśniewski. Questions and inferences. *Logique & Analyse*, 173–175:5–43, 2001.
[54] A. Wiśniewski. Erotetic search scenarios. *Synthese*, 134:389–427, 2003.
[55] A. Wiśniewski. Erotetic search scenarios, problem solving, and deduction. *Logique & Analyse*, 185–188:139–166, 2004.
[56] A. Wiśniewski. Socratic proofs. *Journal of Philosophical Logic*, 33:299–326, 2004.
[57] A. Wiśniewski. Reducibility of safe questions to sets of atomic yes-no questions. In J. Jadecki and J. Paśniczek, editors, *Lvov-Warsaw School: The New Generation*, pages 215–236. Rodopi, Amsterdam/NY, 2006.
[58] A. Wiśniewski. *Essays in Logical Philosophy*. LIT Verlag, 2013.
[59] A. Wiśniewski. *Questions, Inferences, and Scenarios*. College Publications, 2013.
[60] A. Wiśniewski and D. Leszczyńska-Jasion. Inferential erotetic logic meets inquisitive semantics. Research report no. 2(2)/2013, Department of Logic and Cognitive Science, Institute of Psychology, Adam Mickiewicz University, 2013.
[61] A. Wiśniewski and J. Pogonowski. Interrogatives, recursion, and incompleteness. *Journal of Logic and Computation*, 20:1187–1199, 2010.